639.9797/Peo

SAC CRAIBSTON
27 MAR 2006
LIBRARY

About Island Press

Island Press is the only nonprofit organization in the United States whose principal purpose is the publication of books on environmental issues and natural resource management. We provide solutions-oriented information to professionals, public officials, business and community leaders, and concerned citizens who are shaping responses to environmental problems.

In 2004, Island Press celebrates its twentieth anniversary as the leading provider of timely and practical books that take a multidisciplinary approach to critical environmental concerns. Our growing list of titles reflects our commitment to bringing the best of an expanding body of literature to the environmental community throughout North America and the world.

Support for Island Press is provided by the Agua Fund, Brainerd Foundation, Geraldine R. Dodge Foundation, Doris Duke Charitable Foundation, Educational Foundation of America, The Ford Foundation, The George Gund Foundation, The William and Flora Hewlett Foundation, Henry Luce Foundation, The John D. and Catherine T. MacArthur Foundation, The Andrew W. Mellon Foundation, The Curtis and Edith Munson Foundation, National Environmental Trust, National Fish and Wildlife Foundation, The New-Land Foundation, Oak Foundation, The Overbrook Foundation, The David and Lucile Packard Foundation, The Pew Charitable Trusts, The Rockefeller Foundation, The Winslow Foundation, and other generous donors.

The opinions expressed in this book are those of the author(s) and do not necessarily reflect the views of these foundations.

About Defenders of Wildlife

Defenders of Wildlife is dedicated to the protection of all native wild animals and plants in their natural communities. We focus our programs on what scientists consider two of the most serious environmental threats to the planet: (1) the accelerating rate of extinction of species and the associated loss of biological diversity, and (2) habitat alteration and destruction. Long known for our leadership on endangered species issues, Defenders of Wildlife also advocates new approaches to wildlife conservation that will help keep species from becoming endangered. Our programs encourage protection of entire ecosystems and interconnected habitats while protecting predators that serve as indicator species for ecosystem health. Founded in 1947, Defenders of Wildlife is a 501(c)(3) membership organization with over 460,000 members and supporters nationwide.

PEOPLE
and
PREDATORS

PEOPLE *and* PREDATORS

From Conflict to Coexistence

Edited by
Nina Fascione, Aimee Delach,
Martin E. Smith

Foreword by James A. Estes

Defenders of Wildlife

ISLAND PRESS
Washington   Covelo   London

Copyright © 2004 Defenders of Wildlife
All rights reserved under International and Pan-American Copyright Conventions. No part of this book may be reproduced in any form or by any means without permission in writing from the publisher: Island Press, 1718 Connecticut Ave. NW, Suite 300, Washington, DC 20009.
ISLAND PRESS is a trademark of The Center for Resource Economics.

Library of Congress Cataloging-in-Publication data.

People and predators: from conflict to coexistence/ edited by Nina Fascione, Aimee Delach, and Martin E. Smith; foreword by James A. Estes, Defenders of Wildlife.

   p. cm.

Includes bibliographical references and index.

ISBN 1-55963-083-3 (cloth: alk. paper) — ISBN 1-55963-084-1 (pbk.: alk. paper)

   1. Carnivora—Ecology—North America—Congresses. 2. Carnivora—Effect of human beings on—North America—Congresses. 3. Wildlife management—North America—Congresses. I. Fascione, Nina. II. Delach, Aimee. III. Smith, Martin E. (Martin Edgar), 1955-

QL737.C2P36 2004

639.97'97—dc22

2004004597

British Cataloguing-in-Publication data available.

Printed on recycled, acid-free paper ✲

Design by Teresa Bonner

Manufactured in the United States of America

10 9 8 7 6 5 4 3 2 1

Contents

Foreword by James A. Estes   xi
Preface   xv

Introduction   1
  Nina Fascione, Aimee Delach, and Martin E. Smith

PART 1. Coexistence in Rural Landscapes

Chapter 1. Minimizing Carnivore-Livestock Conflict:
The Importance and Process of Research in the Search
for Coexistence   13
  Stewart W. Breck

Chapter 2. Characteristics of Wolf Packs in Wisconsin:
Identification of Traits Influencing Depredation   28
  Adrian P. Wydeven, Adrian Treves, Brian Brost,
  and Jane E. Wiedenhoeft

Chapter 3. Wolves in Rural Agricultural Areas of Western
North America: Conflict and Conservation   51
  Marco Musiani, Tyler Muhly, Carolyn Callaghan,
  C. Cormack Gates, Martin E. Smith, Suzanne Stone,
  and Elisabetta Tosoni

PART 2. Coexistence in Developed Landscapes

Chapter 4. Ecology and Management of Striped Skunks,
Raccoons, and Coyotes in Urban Landscapes   81
Stanley D. Gehrt

Chapter 5. Birds of Prey in Urban Landscapes   105
R. William Mannan and Clint W. Boal

Chapter 6. Challenges in Conservation of the Endangered
San Joaquin Kit Fox   118
*Howard O. Clark Jr., Brian L. Cypher, Gregory D. Warrick,
Patrick A. Kelly, Daniel F. Williams, and David E. Grubbs*

Chapter 7. Carnivore Conservation and Highways:
Understanding the Relationships, Problems, and Solutions   132
Bill Ruediger

Chapter 8. Living with Fierce Creatures? An Overview and
Models of Mammalian Carnivore Conservation   151
David J. Mattson

PART 3. Coexistence in Political Landscapes

Chapter 9. Dispersal and Colonization in the Florida Panther:
Overcoming Landscape Barriers—Biological and Political   179
David S. Maehr

Chapter 10. State Wildlife Governance and
Carnivore Conservation   197
Martin Nie

Chapter 11. Conserving Mountain Lions in a
Changing Landscape   219
Christopher M. Papouchis

Chapter 12. Restoring the Gray Wolf to the Southern
Rocky Mountains: Anatomy of a Campaign to Resolve a
Conservation Issue   240
  Michael K. Phillips, Rob Edward, and Tina Arapkiles

Conclusion   263
  Nina Fascione, Aimee Delach, and Martin E. Smith

  About the Editors   269
  About the Contributors   271
  Index   275

Foreword

Never before in the history of life on earth have global ecosystems been so dominated by a single species. Human influences are everywhere, even in the most remote and wild places. In that sense the word *pristine* is now just a concept, something to be remembered or imagined but never again seen. And along with these changes are worrisome trends—the loss of species and even threats to our planet's life support systems. This is no revelation. Realists have known where we were headed since at least the time of Malthus. The world population presently stands at about 6.3 billion. Forecasters tell us that it will grow to nearly twice that number before it stabilizes or declines, and future threats to biodiversity and global ecosystem services are certain to track the human population. This is the reality we all must face. The goals of those concerned with this reality are to save as many of the pieces as possible, and somehow to hold our planet's ecosystems together through the growth of humanity. The good news is the rise of conservation biology, and that more and more people are dedicating their lives to achieving these goals.

Carnivores are of special interest to conservation biology. These animals, once common nearly everywhere, are now rare or absent in most places. Indeed, the mid- to large-sized carnivores are typically the first species to be lost in human-dominated ecosystems on land and in the sea. Why is this, and what can be done about it? The why is easy. For one, we have selectively killed these creatures for food, or because they eat us and the things we value. For another, the larger carnivores are commonly wide ranging and relatively rare, and thus the maintenance of viable populations typically requires larger tracts of habitat than are needed to maintain viable populations of most other species. Some of the details underlying these generalities are exposed in the following chapters through a variety of case studies, each of which, in one way or another, is built around conflict relationships between humans and carnivorous vertebrates. The reader will see that the conflicts are widespread across rural and developed landscapes, and even across the political fabric of human societies.

The question of what can be done about these conflicts is harder to answer. Laws and regulations clearly are not enough. Habitat losses and resource conflicts are squeezing carnivores out of existence with the inevitable upward march of human numbers. Compromise is not the answer. We've already lost far too much. Sequestering these creatures in parks and refuges is not the answer. There are simply too many species with too many differing needs to manage that at this late time. Besides, who wants to live in a world in which all of wild nature is stuffed away in a park somewhere? Coexistence, the theme of this volume, is the only possible solution. But can we possibly coexist with carnivores, or more properly stated, can they coexist with us? Probably not, so long as we view them as our enemies and believe that living with them incurs costs without conveying benefits. If, on the other hand, there were some broadly perceived benefit to living with carnivores that outweighed the cost, more people just might be willing to adopt coexistence as a realistic goal.

In fact, there are benefits to living with carnivores that few yet realize. The problem is that the costs are simple and tangible, whereas the benefits are complex and not immediately expressed in monetary terms. A wolf eats the rancher's cow. That's a simple and tangible cost to living with wolves. But wolves also eat elk, or scare them away from places they might otherwise roam and browse at will. And elk eat trees, which in turn stabilize hillsides and maintain riparian and floodplain vegetation. And birds and other wildlife utilize this vegetation, as do people in a variety of direct and indirect ways. Now the imbalance of costs and benefits to living with wolves isn't so clear.

A sea otter eats the fisher's abalone. That's a simple and tangible cost to living with sea otters. But sea otters also eat sea urchins, which in turn eat kelp. And without kelp many fish are reduced, and waves strike the shoreline with greater force than they would if the kelp were abundant, and many other things change that we humans value. Now the imbalance of costs and benefits to living with sea otters isn't so clear. And so it probably is for one carnivore species after the next.

Species interact with one another in complex ways that are difficult for us to see and understand. Most of these stories have never been told because most are still undiscovered. Might it be that such benefits could persuade more people to forego the costs of living with carnivores? While that seems like a stretch, it may also be the best and most utilitarian hope there is for carnivore conservation.

We stand at a point in time when wild things and wild places are disappearing rapidly. Large carnivores are perhaps the most poignant symbol of these losses. But that emotional rendering has not been sufficient to redirect policy, nor does it capture the enormity of all that we are losing. Our perception of the loss of species is reasonably accurate, while our understanding of the associated loss of species interactions—the complex effects of wolves on terrestrial ecosystems or sea otters on kelp forests—is miniscule almost beyond imagination. The challenge is to open our minds, to learn about these interactions, and to imagine what might be. With that knowledge lies hope and opportunity: a reason not to lock up our carnivores in a park somewhere and pray that some minimalistic vision of a viable population will suffice in preserving them for future generations; a pathway by which we might indeed effect a transition from conflict to coexistence between people and predators.

J. A. Estes, U.S. Geological Survey
University of California, Santa Cruz

Preface

On November 16–20, 2002, more than 800 scientists, activists, and educators came together in Monterey, California, for a three-day conference on predators called "Carnivores 2002: From the Mountains to the Sea." This was the fourth in a series of biennial carnivore conferences hosted by Defenders of Wildlife, a national, nonprofit conservation organization dedicated to the preservation of native species and the habitats on which they depend. The Monterey meeting had a broad focus that included not only large terrestrial carnivores but mesocarnivores, marine carnivores, and raptors. There were more than 150 speakers and 50 posters covering a range of topics from ecology to management, behavior to threats, and research tools to human dimensions. The enthusiasm generated by the meeting led to discussions of a book based on selected talks from the conference, focusing on the interface between humans and predators. Thus *People and Predators: From Conflict to Coexistence* was born.

Given the breadth and variety of topics addressed at "Carnivores 2002," the conference could have provided enough material for ten books about predators. However, for the purpose of creating a volume that would have wide appeal and applicability, the authors chose to focus on conflicts between carnivores and humans: the causes, possible solutions, and the relevance of conflict and resolution to the successful persistence of carnivore populations. Finding ways to successfully resolve issues that occur when humans and carnivores overlap in habit and habitat has been a crucial need in the recovery and conservation of wolves, bears, otters, and other species. This need will only grow as the popular and legal mandates to conserve and restore biodiversity, particularly carnivore species, intersect with trends of human population growth, suburbanization, and changing land use patterns.

Each chapter of *People and Predators* deals with a facet of the interactions between carnivores and humans, provides background on a particular problem, and includes a case study describing how these challenges have been met, or what research or tools are still needed to resolve the conflict at hand. The species and conflicts described reflect much of the ground covered at "Carnivores 2002"; however, owing to constraints of time and space, *People and Predators* is focused almost exclusively on North American terrestrial carnivores. Thus, this volume is by no means a comprehensive treatment of possible conflicts and solutions, or even fully representative of those issues raised at "Carnivores 2002." We believe, however, that this volume provides concrete tools for the resolution of many of the problems that stand between us and a future in which carnivores fulfill their historic ecological roles.

As an outgrowth of the "Carnivores 2002: From the Mountains to the Sea" conference, this book could never have come to pass without the tireless efforts of all those who assisted with the conference. We are fortunate to work with such a dedicated group of conservationists at Defenders of Wildlife. We also wish to thank the staff of the Monterey DoubleTree Hotel; the Monterey Convention Center, where the conference was held; and the Monterey Bay Aquarium, which graciously donated space for our "icebreaker" event. We also acknowledge the participants and presenters, without whom the conference could not have happened, particularly our keynote speakers: Sylvia Earle, James Estes, and George Rabb. We would especially like to thank Rodger Schlickeisen and Mark Shaffer for their leadership on Defenders of Wildlife's carnivore conservation efforts.

The numerous contributors to this book were a pleasure to work with. We thank them all for their research, their insights, and their enthusiastic cooperation in this project. We are very grateful to Terry Pelster, Sharon Wilcox, J. Christopher Haney, and Miriam Stein for their help with manuscript preparation. Barbara Dean and Laura Carrithers, Jessica Poppe, and Brighid Willson at Island Press were instrumental in guiding this process from conference to completion, and for that, we are deeply appreciative. And most of all, the authors would like to thank our spouses, Stephen Kendrot, Robert Barber-Delach, and Hanne Hansen, for their constant support and encouragement.

Introduction

Nina Fascione, Aimee Delach, and Martin E. Smith

Throughout the centuries, predators have always held a unique place in the human psyche, inspiring both fear and awe (Kellert et al. 1996). Images of carnivores are diverse, but whether the emotions are positive or negative, carnivores fill our imagination in ways that are larger than life. For some people, carnivores elicit fears of child-snatching, bloodthirsty killers; others see heroic images used as symbols of cultural traditions, emblems in sport, and powerful automobiles. Carnivores also play a unique role in our ecosystems, serving as keystone species that help regulate the environment around them in beneficial ways. Yet predators can have more tangible, sometimes detrimental, impacts on humans. Wolves (*Canis lupus*) and grizzly bears (*Ursus arctos*), for example, do occasionally kill livestock, and grizzly bears and mountain lions (*Puma concolor*) have injured and killed humans in North America. Urban carnivores such as raccoons (*Procyon lotor*) can be considered a nuisance or transmit disease. So a basic question remains: How do we manage the world's carnivore populations to conserve this important natural resource while mitigating any harmful impacts?

People and Predators: From Conflict to Coexistence examines these complex human-carnivore relationships and investigates how humans can work to preserve this group of animals while protecting human lives and livelihoods. The goal of this volume is not only to highlight some of the problems encountered when humans and predators live in close proximity, but to offer some proven and potential solutions to such problems, so that we can conserve predators while maintaining a harmonious coexistence well into the future. Our desire is to help wildlife professionals and the general public better understand a few of the issues facing carnivore conservation in the twenty-first century, and to provide tools for dealing with those problems.

No one book can adequately address the full range of carnivores and the variety of challenges that sharing the planet with them might entail. The chapters contained in this volume are examples of the various issues, and many of the solutions are pertinent to management of other taxa. Though the experience of living with carnivores is certainly a global one, *People and Predators: From Conflict to Coexistence* focuses on North America, since there is an almost overwhelming array of issues in this region alone. This continent has demonstrated the full gamut of human-carnivore relationships over the centuries, from early Native American use of predators as totems to the modern-day problem of nuisance urban carnivores. As the human population in North America continues to expand and encroach on landscapes that historically belonged to wildlife, it will become more important than ever to move from conflict to coexistence.

Living with Predators

It might be fair to wonder why we bother living with predators at all. If they eat our livestock, threaten our safety, and compete for desired resources, perhaps it makes sense to eliminate this group of animals entirely. Fortunately, predators also provide innumerable benefits that make it worthwhile to develop methods of harmonious coexistence. Carnivores help maintain ecosystem function by acting as keystone species, helping to regulate the surrounding environment (Berger 1999; Terborgh et al. 1999; Miller et al. 2001). Economists are discovering that carnivores can also be good for the bottom line. Wolf restoration in the northern Rockies has stimulated a tremendous growth in wildlife-related tourist dollars. Current studies will determine the exact nature of this wolf-driven economy, but estimates taken just prior to the restoration estimated annual income of $20 million (Duffield 1992).

Similarly, recovery of California sea otters (*Enhydra lutris*) is expected to generate tourism income of $176 million annually and to create between 1,400 to 8,400 jobs supported by tourism dollars. On top of that, ecosystem services associated with sea otter maintenance of healthy kelp forests would amount to $7,600 per acre per year (Loomis, in prep.).

Many would argue that carnivores provide a rich cultural and historical heritage that should be maintained. It cannot be denied that many North Americans derive an emotional value just by knowing predators exist (Kellert et al. 1996).

Today our knowledge of the role predators occupy in the ecosystem is greater than ever. In a system known as the "trophic cascade," which describes the interactions between different levels of the food chain, predators exert an influence on more than just the numbers of their direct prey. Perhaps the most classic example of a keystone species is seen with the southern, or California, sea otter, which acts as a steward of the Pacific marine environment. By preying on kelp-eating sea urchins (*Strongylocentrotus puratus*) and other marine herbivores, sea otters protect kelp and other macro algae, allowing them to flourish and provide habitat for numerous species of fish. If sea otters are removed or diminished, these rich ecosystems collapse as sea urchins, unchecked, devour the kelp beds and destroy habitat (Estes and Palmisano 1974).

A more recent example of the keystone predator effect has been seen in Yellowstone National Park since the reintroduction of wolves in 1995 and 1996. While tourists flock to Yellowstone's Lamar Valley to witness for themselves one of the park's most glamorous predators, forest researchers are equally fascinated by the gradual return of quaking aspen, one of the most ecologically important riverside tree species in the Yellowstone ecosystem. Scientists have determined that there is a direct link between the recovery of the ecosystem's top predator and the trees. Aspen growth essentially stopped once wolves were removed from the park in the 1920s. Now that the wolves have returned, the trees are growing again (Ripple and Larsen 2000; Ripple et al. 2001).

Elk, as it turns out, forage differently depending on whether predators are present. During the 60 years wolves were absent from the park, elk spent more time browsing alongside rivers, trampling the low vegetation and inhibiting new growth of native tree species, including aspen. With wolves once again on the scene, elk are behaving more cautiously—avoiding areas with dense foliage and spending more time in open areas to keep an eye on their surroundings. This behavior change is altering the entire landscape. By altering the movements and foraging patterns of elk, wolves are playing a key role in preserving the integrity of Yellowstone's overall biodiversity.

As aspen and the other riparian vegetation grow taller and expand in canopy cover, the beneficial impacts to the system will include stream channel stabilization, floodplain restoration, and higher water tables. Through a tropic cascade effect that improves riparian habitat, wolves may be beneficial to numerous species of vertebrates and invertebrates, including fish, birds, beaver (*Castor canadensis*), and butterflies, as well as many other species.

There are other beneficial chain reactions stemming from wolves' return to Yellowstone. Coyotes (*Canis latrans*), overpopulated for years in the absence of the larger canid, have been reduced by up to 50% in some regions, which in turn has enabled smaller animals such as foxes (*Vulpes vulpes*) and rodents to rebound (Crabtree and Sheldon 1999). With more rodents available, birds of prey have thrived. Leftover wolf kills are providing a reliable, year-round food source for a plethora of other species, including ravens (*Corvus corax*), magpies (*Pica hudsonia*), golden and bald eagles (*Aquila chrysaetos* and *Haliaetus leucocephalus*), foxes, cougars, insects, and the park's famous grizzly bears. It would be hard to argue that the return of this top predator has not been beneficial.

Many carnivores are also labeled "umbrella" species. Protecting these predators and their habitats will, by default, preserve other, generally smaller, species that rely on the same ecosystems. Grizzly bears in particular are considered an umbrella species. Because they are large and wide ranging, conservation measures for grizzlies help protect a host of other species. Finally, many predators are also known as "flagship" species. Because they are often considered to be attractive or "sexy," they generate public interest and funds in protecting habitats that benefit other species as well.

Living with Humans

There is one carnivore that exerts a stronger influence on our ecosystems than all the rest—human beings. Though we tend to hear more about the threats predators pose to us, the fact is that we have long shaped the environment with our actions, and human activities have tremendous impacts—most of them deleterious—on predators. As human populations grow and wilderness areas shrink, it will become ever more imperative to find means to make room for predators, and methods of coexistence.

The global human population now exceeds 6 billion and is growing. Though our earth is large, we are consuming natural resources at an unprecedented rate and, at least in North America, apparently more rapidly than nature is able to replenish itself (Commission for Environmental Cooperation 2001). With less than 10% of the Earth's land surface protected in some form (World Conservation Union 2000), there are increasingly fewer wild places remaining for wildlife. Additionally, some parks are protected on paper only, lacking the resources to manage wildlife for continued viability. Experts estimate that we will lose between 1 and 5 million species of plants and animals in the next few decades, primarily owing to habitat destruction, a pace of extinction that is as much as 10,000 times the historical background rate (Wilson 1989). If present trends continue, half of all the species on Earth may be on the road to extinction within 100 years.

Carnivores are particularly vulnerable, as many are wide ranging and require larger tracts of relatively undeveloped habitats in which to live. Other indirect human factors such as depletion of natural prey, pollution, and artificially introduced species and their accompanying diseases, as well as direct human persecution in the form of hunting and commercial trade, combine to threaten the very existence of many carnivore species (Weber and Rabinowitz 1996). The American jaguar (*Panthera onca*) of the southwestern United States and northern Mexico is dwindling. Mountain lions have been extirpated from most of the east. Lynx (*Lynx canadensis*) populations are depleted throughout most of their historical range. The grizzly bear is listed as threatened under the Endangered Species Act (ESA) and exists in only five small populations in the lower forty-eight states. The southern sea otter is imperiled owing to conflicts with humans, primarily illegal killing for perceived competition over resources, and disease, probably introduced by humans. Population counts of mesocarnivores such as wolverine (*Gulo gulo*), fisher (*Martes pennanti*), and badger (*Taxidea taxus*) are mostly estimates, since they are a difficult group to study, but all indications are that many populations have declined.

Fortunately, not all the news regarding carnivores in North America is bad. The public has a tremendous interest in predators and, for the most part, cares deeply about these species. This was demonstrated in no uncertain terms during the federal comment period on wolf restoration in the northern Rockies. More than 160,000 comments were sent by the public in 1993 on the final environmental impact statement (E. Bangs, pers. comm.) on the reintroduction of gray wolves to Yellowstone National Park and central Idaho, one of the highest number of public comments received to date on any wildlife issue. As a result of that restoration project and protection of wolves in the Great Lakes states, wolf populations in the contiguous United States have increased from fewer than 1,000 in the mid-1970s to roughly 4,000 today. Similarly, Endangered Species Act protection and a ban on the pesticide DDT are responsible for the recovery of bald eagles and peregrine falcons. And, though still on a perilous road to recovery, the panther population in south Florida is hovering at around 100 after being down to a handful of individuals prior to restoration efforts.

The key question facing wildlife managers and legislators is how to manage rare, as well as common, carnivores while addressing the needs of both predators and people. The key questions facing society are whether we will make room for predators and whether we will tolerate them. There are myriad issues challenging a peaceful coexistence between humans and predators. These dilemmas can be classified generally as "ways in which predators threaten humans and our livelihoods" and "ways in which humans threaten predators and their livelihoods." In other words, coexistence is a two-way street. Wolves eating sheep would clearly fall under the first category, and humans building roads in sensitive habitat would fall into the second group. Of equal importance are the ways in which humans react to predators and the level of value placed on their conservation.

Another way to examine the issue is through the various landscapes in which we coexist with predators. In *People and Predators: From Conflict to Coexistence,* we divide these landscapes into rural, developed, and political. The three chapters in Part 1, "Coexistence in Rural Landscapes," discuss the challenges of maintaining predator populations in rural areas. Through case studies from three regions—the Great Lakes, the northern Rockies, and western Canada—our authors ascertain the damage large carnivores can inflict on farming and ranching interests by preying on livestock. The authors also propose solutions to predation problems through innovative preventative technologies and livestock management practices that reduce, and in some cases help eliminate, livestock depredation.

Part 2, "Coexistence in Developed Landscapes," examines how the human-carnivore relationship changes as the landscape becomes more developed. In these landscapes, the conflicts are diverse and include problems caused by predators, as well as problems caused by humans. The first obvious challenge in a developed landscape is maintaining enough habitat to support viable predator populations. However, many carnivores can survive in developed landscapes, and residing in such close quarters to humans can provide for a wide variety of conflicts. Our authors examine such issues as how humans can share the landscape with mid-sized terrestrial carnivores and urban birds of prey; how human-introduced invasive species can negatively impact native carnivores; how designing wildlife corridors can greatly reduce the negative impacts of habitat fragmentation; and how we can conserve declining species despite human impacts.

Part 3, "Coexistence in Political Landscapes," offers insight into some of the sociopolitical factors impacting carnivore conservation. The challenges faced in the political, legal, and economic arenas may be the toughest to overcome. The four chapters in this final section provide case studies of management challenges for wolves and mountain lions, and suggest recipes for solutions.

Clearly, more research and discussion are necessary for finding additional solutions to these global, complex challenges. In addition to funds, solving these issues will require all-inclusive stakeholder input, dissemination of research findings through education and outreach, and constant evaluation of our ethical responsibilities. Because technology continues to improve and new technologies are continually developed, there will always be a need to analyze and evaluate the effectiveness of new tools for managing large predators. Additionally, cultural values and ethics are constantly changing, triggering a need to frequently examine our goals and desires for conserving carnivores.

As human populations expand and we alter our environment through the growth of our own species and our consumption patterns, our decisions may become more difficult. Can we find a way to make room for both humans and carnivores, particularly the large predators, in the next century? Though they are complex and often a management challenge, carnivores serve a key role in North American ecosystems and cultural heritage, and both humans and predators deserve to share our continent's wildlands, as well as our rural farmlands and urban parks. What we hope to provide with *People and Predators* is an insight into the variety of challenges, and practical solutions to some common problems. We also want to promote a continuing discussion because it is important that this issue be examined not only by wildlife professionals but by an interested public, because moving from conflict to coexistence will take an effort from all of us.

Literature Cited

Berger, J. 1999. "Anthropogenic extinction of top carnivores and interspecific animal behaviour: Implications of the rapid decoupling of a web involving wolves, bears, moose and ravens," *Proceedings of the Royal Society of London*, Series B, 266:2261–2267.

Commission for Environmental Cooperation. 2001. *The North American Mosaic: A State of the Environment Report.* www.cec.org/files/PDF/PUBLICATIONS/soe_en.pdf.

Crabtree, R. L., and J. W. Sheldon. 1999. "Coyotes and canid coexistence," in *Carnivores in Ecosystems: The Yellowstone Experience,* ed. R. W. Clark, A. P. Curlee, S. C. Minta, and P. M. Kareiva, 127–163. Yale University Press, New Haven, CT.

Duffield, J. W. 1992. "An economic analysis of wolf recovery in Yellowstone: Park visitor attitudes and values," in *Wolves for Yellowstone? A Report to the United States Congress,* ed. J. D. Varley and W. G. Brewster. Vol. 4, Research and Analysis. National Park Service, July.

Estes, J. A., and J. F. Palmisano. 1974. "Sea otters: Their role in structuring nearshore communities," *Science* 185:1058–1060.

Kellert, S. R., M. Black, C. R. Rush, and A. J. Bath. 1996. "Human culture and large carnivore conservation in North America," *Conservation Biology* 10:977–990.

Loomis, J. In prep. "Economic benefits to California residents of southern sea otter." Report to Defenders of Wildlife.

Miller, B., B. Dugelby, D. Foreman, C. Martinez del Rio, R. Noss, M. Phillips, R. Reading, M. Soulé, J. Terborgh, and L. Willcox. 2001. "The importance of large carnivores to healthy ecosystems," *Endangered Species Update* 18:202–210.

Ripple, W. J., and E. J. Larsen. 2000. "Historic aspen recruitment, elk, and wolves in northern Yellowstone National Park, USA," *Biological Conservation* 95:361–370.

Ripple, W. J., E. J. Larsen, R. A. Renkin, and D. W. Smith. 2001. "Trophic cascades among wolves, elk and aspen on Yellowstone National Park's northern range," *Biological Conservation* 102:227–234.

Terborgh, J., J. A. Estes, P. C. Paquet, K. Ralls, D. Boyd-Heger, B. Miller, and R. Noss. 1999. "The role of top carnivores in regulating terrestrial ecosystems," in *Continental Conservation: Scientific Foundations of Regional Reserve Networks,* ed. M. E. Soulé and J. Terborgh, 60–103. Island Press, Washington, DC.

Weber W., and A. Rabinowitz. 1996. "A global perspective on large carnivore conservation," *Conservation Biology* 10:1046–1055.

Wilson, E. O. 1989. "Threats to Biodiversity," *Scientific American* 261:108–116.

World Conservation Union. 2000. *Protected Areas: Benefits Beyond Boundaries.* www.iucn.org/themes/wcpa/pubs/pdfs/WCPAInAction.pdf.

PART 1

Coexistence in Rural Landscapes

Humans are moving into previously undisturbed habitat at an ever-quickening rate, and thus conflicts with predators and disruptions of their prey base are on the rise. The increase in encounters coincides with a greater awareness by the general public of the ecological role that predators play in preserving healthy ecosystems and thus increases the demand that predators be protected and preserved. On the front line of this human-predator interface are the many rural residents who still make their livelihood from farming and ranching. These rural residents sometimes have to suffer the true economic burden of "living with carnivores" as they lose livestock to depredation.

Historically, carnivore management has meant simply shooting the predators until the "problem" was extirpated. This approach is no longer acceptable. Socially, the political power has shifted somewhat to the more urban society, whose members analyze the conflicts from afar and logically determine that preserving wolves or grizzly bears has intrinsic value. In addition, the maturing science of wildlife biology has begun to quantify the negative ramifications of not preserving the top predators within an ecosystem. And yet ranchers still need to make a living from the land. Modern wildlife managers understand both of these perspectives and strive to find solutions that will allow predators and ranchers to coexist. Part 1 of this book describes conflicts in rural landscapes in more detail and outlines several ways in which researchers and managers are striving to resolve the conflicts.

The chapters in Part 1 address the rural landscape, an area that is important regardless of where we live, because most of our food (crops and livestock) is grown there and most of our remaining natural areas are located there. Owing to their connection to the land, some rural residents have a keen sense of the ecological interrelationships within a functioning ecosystem, because without such knowledge crops fail and livestock perish. Some rural residents also work directly with animals on a daily basis, resulting in an added insight into many aspects of animal behavior. It is also in these rural regions that carnivores and humans first make contact and obviously where conflicts first occur. Therein lies the problem. The coexistence of humans with large predators is one of modern wildlife management's most challenging undertakings and one that promises to become more difficult in the years ahead.

Chapter 1, "Minimizing Carnivore-Livestock Conflict: The Importance and Process of Research in the Search for Coexistence," by Dr. Stewart Breck, explores the importance of research in the effort to develop solutions for conflicts between large carnivores and livestock. Dr. Breck explores a number of different research topics that will enhance our ability to predict and mitigate carnivore-livestock conflict.

Chapter 2, "Characteristics of Wolf Packs in Wisconsin: Identification of Traits Influencing Depredation," is a collaboration by a group of agency, academic, and independent scientists led by Dr. Adrian Wydeven. In this chapter the authors follow the depredation trends of a naturally recovering wolf population and describe a case study from the state of Wisconsin that illustrates how managers and researchers might use predisposing factors to predict and avert wolf attacks on domestic animals.

Chapter 3, "Wolves in Rural Agricultural Areas of Western North America: Conflict and Conservation," is by a group of scientists from academia and nongovernmental organizations, led by Dr. Marco Musiani of the University of Calgary. These researchers review wolf-livestock conflicts, and the recovery, conservation, and legal status of wolves, including compensation and depredation management for Alberta, Idaho, Montana, and Wyoming. Understanding the factors that influence depredation allows us to consider alternative approaches for managing depredation risk while encouraging wolf conservation.

Together these chapters provide the reader with an insight into the complexity of coexisting with carnivores in the rural landscape. We make no attempt to present a complete representation of all the research and management projects underway throughout North America and, indeed, throughout the world. Rather, these chapters offer some idea of the breadth of research and management solutions being attempted in different areas of North America. The reader should also come away with a better understanding of how intensely the problem is felt by those humans who earn their livelihood from utilizing rural regions of the world. It is our hope that a mutual respect and tolerance will be developed by both rural and urban residents such that progress can continue to be made toward coexisting with carnivores.

CHAPTER 1

Minimizing Carnivore-Livestock Conflict: The Importance and Process of Research in the Search for Coexistence

Stewart W. Breck

Large carnivores that kill livestock can create both an economic and an emotional burden for producers. Because these acts of predation often result in deep disdain for large carnivores, and because humans who are in the profession of producing food have strong local and national political backing, finding solutions that minimize the number of livestock killed by large carnivores is a critical conservation priority. This is particularly true in areas throughout the world where the repatriation and protection of large carnivores is a priority. Unfortunately, our understanding of the biological and ecological factors that influence a carnivore's decision to kill livestock, and of alternative methods for mitigating problems, is still nascent. Attaining solutions that minimize livestock losses while maintaining carnivore populations that are sustainable and ecologically relevant (Pyare and Berger 2003) will require reliable knowledge that only well-designed and well-executed research can provide.

In this chapter I explore the importance of research in the continued effort to develop better solutions for problems between large carnivores and livestock. I begin with a brief overview of research design to help nonscientists understand the terminology of the field. I then explore a number of different research topics, from causative factors in carnivore-livestock conflict to possible ways to reduce such conflict. Each research topic stresses varying needs, such as improving research design; broadening the scope of study to include large-carnivore behavior and ecology as these relate to livestock production; and addressing how environmental characteristics affect our management tools. These are difficult arenas for research because of the inherent problems in working with carnivores, the sociopolitical aspects of carnivore-livestock conflicts, and the large scale at which researchers must ask questions. In conclusion, I suggest ways to improve the scientific process by promoting better cooperation among stakeholders, more collaboration among scientists, and the use of adaptive management.

Research Design

Research may be defined as a purposeful endeavor to attain reliable knowledge—that is, information leading to a trustworthy understanding of a phenomenon, and/or which can be used to make good decisions regarding alteration of a system to achieve a desired outcome. Often these goals involve finding cause-and-effect relationships among variables (e.g., a scare device causes carnivore behavior to be altered). Trust in the reliability of results in large part depends on how well three elements of research design (control, randomization, and replication) are incorporated into the study.

The research design that offers the most opportunity to gain reliable knowledge regarding cause-and-effect relationships is the manipulative experiment. Manipulative experiments purposefully manipulate treatment units and compare them with unmanipulated control units, randomly selecting which units will act as treatment or control, and replicating the study either internally (replication within a study) or externally (replication of an entire study).

Unfortunately, manipulative experiments are extremely difficult to perform with large endangered carnivores. In particular, the lethal removal of individuals (a manipulation) is hard to justify because of ethical concerns and practical considerations regarding species recovery. Working at large spatial scales makes manipulative experiments more difficult for many reasons, including high cost (limiting the ability to replicate); logistical difficulty (limiting the ability to set up controls); and increased environmental variability (increasing the need for replication and control). Because of these and other difficulties, wildlife researchers often perform observational studies, which do not contain manipulations and often lack critical features of sound research design. Attaining reliable knowledge is possible with observational studies, but the process takes longer and is risky in the sense that unreliable information may be attained.

The challenge then is for researchers to maximize efficiency of learning and reliability of knowledge by creatively incorporating as many features of good research design as possible. Conceptually, a great deal of progress has been made along this front by promoting such ideas as adaptive management ("learning by doing"; MacNab 1983; Walters 1997) and metareplication (external replication; Johnson 2002), topics that will be covered in more detail toward the end of the chapter.

Understanding Carnivore-Livestock Conflicts

To develop effective solutions to carnivore-livestock conflicts we must understand why large carnivores attack livestock by discerning what aspects of the predators' environment, biology, or ecology predispose them to do so. These questions require synthesizing knowledge of individual carnivore behavior, population dynamics of carnivores and their prey, and community ecology as they relate to the presence of livestock. Two research areas that begin to address these questions deal with the presence of "hot spots" and the existence of problem individuals.

Hot Spots: Why Do They Exist?

Recent studies of livestock losses to carnivores demonstrate the presence of hot spots, or small areas that have recurring attacks on livestock by carnivores (Fritts 1982; Cozza et al. 1996). For example, Stahl et al. (2001) studied lynx attacks on sheep in France and found that certain geographical areas, covering only 0.3–4.5% of the total area where attacks occurred, accounted for 33–69% of the attacks. It is unclear how ubiquitous hot spots are across carnivore species and across different ranges, and it is not clear what causes hot spots, but a better understanding of this phenomenon may be important for more effective management. A number of factors are hypothesized for causing hot spots, including individual problem predators, herding techniques, the abundance and availability of wild and domestic prey, habitat characteristics, and the abundance of predators.

Research regarding hot spots should first focus on determining their prevalence and distribution for various carnivore species in a diversity of areas. Important data are spatial location of available livestock, number of kills in particular areas, spatial location of kills, and species of carnivore responsible for kills. Attaining such data requires the skill of trained experts who are able to determine the difference between animals killed or scavenged, and to distinguish species responsible for kills. Simultaneously, personnel should collect information on site-specific characteristics (e.g., proximity to dwellings); husbandry practices (e.g., presence or absence of guard animals and methods of carcass disposal); and landscape-level characteristics (e.g., travel corridors, available cover).

Once hot spots have been identified, the next step is to determine what causes them. Mech et al. (2000) performed an analysis along these lines by comparing Minnesota farms that experience chronic wolf depredation of cattle with nearby farms without chronic problems. Of 11 farm characteristics measured, these researchers found that chronic losses occurred on larger farms, farms that had more cattle, and farms that had herds farther from human dwellings. Whether these are general patterns that hold true for other geographical areas and other species of carnivores is unknown, but further study is warranted.

Once site characteristics have been modeled for hot spots, the second step is to validate these models by performing manipulative experiments specifically designed to test each of the factors identified in the model. For example, if it is true that farms with herds farther from human dwellings experience greater losses to wolves (Mech et al. 2000), then experimentally reducing the distance of herds from human dwellings should decrease problems with wolves. To date no such manipulative studies have been performed, which may be a function of the difficulty of performing them and/or a poor understanding of which possible factors to incorporate into such a manipulative experiment.

Problem Individuals

Problem animals are those individuals that kill more livestock per encounter than other individuals within the population. Problem animals are known to exist for a wide range of carnivores, including grizzly bears (*Ursus arctos*; Anderson et al. 2002); coyotes (*Canis latrans*; Till and Knowlton 1983; Conner et al. 1998; Sacks et al. 1999a); lynx (*Lynx canadensis*; Stahl et al. 2002); wolverine (*Gulo gulo;* Landa et al. 1999); and mountain lions (*Puma concolor;* Rabinowitz 1986). However the causal mechanism leading to the development of problem animals is unknown.

Linnell et al. (1999) hypothesize that the development of problem individuals is primarily behavioral in nature and will occur most frequently in situations in which livestock are most intensively managed (i.e., where nonlethal tools and livestock husbandry are commonly practiced). In these areas predators must learn specialized behaviors to successfully kill livestock, which leads to a negative feedback cycle as these individuals become better at overcoming management strategies. In areas where livestock are free ranging and unattended, most individual carnivores have an equal opportunity to kill livestock, and there is no perceptual difference between natural ungulates and livestock. Most predators will kill some livestock, but no one individual will develop specialized behavior and become a problem animal (Linnell et al. 1999).

Other competing hypotheses are more simplistic, implicating adult and/or subadult males, dispersing juveniles, or territorial breeding pairs as individuals with high probabilities of becoming problem animals (Rabinowitz 1986; Sacks et al. 1999b). Finally, learning between mother and offspring and between adults in social species may play a role in the development of problem animals (Breck, pers. observ.).

These hypotheses regarding problem animals have only begun to be scrutinized systematically under controlled conditions. More research is needed to determine how ubiquitous problem individuals are within various populations and various species of carnivores. If research finds that problem individuals are common, then other questions become important, such as, How do problem individuals develop? What characteristics of the environment encourage development of problem animals? and How can we identify and manage problem animals?

Developing effective research programs that can answer questions regarding problem individuals will be very difficult, because the causal mechanisms involved in the development of livestock-killing behavior are probably complex and variable. Depending on the hypotheses being tested, information needs include detailed data on the types of prey selected by individual carnivores, knowledge of the social status of individuals, and knowledge of animal lineages. Attaining such information may prove to be very difficult because large carnivores are secretive, exist in low densities, and are hard to study. New technologies that may help answer questions regarding problem animals include sophisticated radiotelemetry and global positioning system (GPS) technology, which can reveal fine-scale movement and activity patterns (Anderson and Lindzey 2003), as well as molecular technologies that allow the identification of individual animals through DNA analysis of feces, tissue, and saliva.

But new technology cannot make up for poorly designed studies. Performing experiments with adequate replication and control will be necessary, though challenging, if we are to understand what causes the development of problem animals. Depending on the hypotheses, it may be best to first address questions in a "laboratory" setting (e.g., the USDA Wildlife Services outdoor coyote facility in Logan, Utah, and the captive wolf facilities of the Wildlife Science Center in Forest Lake, Minnesota). These captive facilities offer a greater ability to enhance research design by implementing experimental protocols and by increasing replications of a study in a controlled setting. When feasible, results learned in captive facilities should then be tested in field settings with well-designed experiments.

Research to Improve Management

Understanding what influences carnivores to attack and kill livestock will not only provide basic knowledge regarding the phenomena but also aid the development of tools and techniques that managers can use to mitigate problems. It is likely that the most significant advances will unite knowledge of livestock husbandry, technology, and carnivore behavior and ecology. But uniting these disparate fields into coherent and meaningful research studies that develop effective solutions will rely on careful consideration of research design and strong commitment by a variety of stakeholders to the process of research.

Improving Livestock Husbandry

For this chapter, *livestock husbandry* refers specifically to the movement and management of livestock to reduce the number killed by predators. Under this definition, common techniques of livestock husbandry include proper disposal of carcasses to prevent attracting predators; managing birthing dates so young are not born on the open range; herding vulnerable animals at night; and combining herds so livestock are not spread evenly across the landscape. In areas where large carnivore populations were extirpated, husbandry techniques (herding in particular) have been abandoned primarily because of the high cost of labor. With the reestablishment of large carnivore populations in areas throughout the world, better understanding of husbandry will be critical as the various techniques may offer effective and economically sound long-term solutions for reducing conflict in areas where livestock are free ranging.

Robel et al. (1981) evaluated the effectiveness of several husbandry methods for reducing sheep losses to coyotes by correlating the number of sheep killed to a number of factors that varied among 109 sheep producers in Kansas. Producers experienced less predation loss when they hauled away sheep carcasses, lambed during particular seasons, confined flocks of sheep to corrals, and maintained larger flock sizes. Evidence from Europe also suggests the importance of husbandry. Greater losses of livestock to carnivores occurred in Norway, where sheep were entirely free ranging and unattended, as compared with France, where livestock were constantly herded or confined at night (Stahl et al. 2001). Though these and other studies suggest husbandry can be effective for reducing conflict with carnivores, our knowledge regarding husbandry and its effectiveness with different carnivore species is still very limited (Knowlton et al. 1999).

Observational studies of kill rates on livestock in a variety of habitat types and under a variety of husbandry levels will be helpful in quantifying the problem across a range of carnivores and variety of habitats. However, because kill rates on livestock can vary spatially and temporally, combining the results of these studies to make inferences regarding husbandry techniques may be difficult to interpret. To give a better indication of the influence of husbandry on kill rates, experimental studies are needed that control for temporal and spatial variability.

An ideal study would compare carnivore kill rates on livestock from a treatment site using designated husbandry practices with a control site not utilizing such husbandry practices. Both sites would have similar wild and domestic ungulate densities, carnivore densities, and environmental characteristics. The general design of this study would then be repeated in more than one area to increase confidence in the results. Implementing such a study may be logistically impossible, but understanding the components that lead to an ideal study design can help when making decisions where one must compromise study design. It may be that husbandry practices (i.e., the treatment) could be applied at the same site and that control data (e.g., kill rates of livestock in the absence of herding) could be collected for a period of several years before and/or after treatment data (e.g., kill rates of livestock in the presence of herding).

Regardless of the final design, successful completion of an experimental study regarding husbandry practices will require tremendous commitment to learning by a variety of stakeholders. In particular, producers whose livestock are part of the study will have a great deal vested in terms of financial and emotional commitment. It is essential to a research project that producers understand the importance of attaining baseline or control data and are willing to tolerate losses of livestock to carnivores without altering practices or deviating from the study protocol. To encourage commitment from the producer, it is essential that good communication is established between researchers and the producer and that compensation be provided for all livestock lost to carnivores during the study.

Tools for Managing Carnivores

Tools for managing carnivores include exclusionary items (e.g., scare devices, dogs, and fences); devices to capture carnivores (e.g., snares and leghold traps); drugs to inhibit reproduction (e.g., porcine zona pellucida [PZP], mifepristone, and cabergoline); supplemental feed (e.g., high-protein sweat feed); and devices to lethally control individual problem animals (e.g., livestock protection collars).

Developing tools that are efficient, reliable, and useful requires two types of knowledge: methodological and field. Methodological knowledge is information about a tool. Procurement of this knowledge generally occurs in settings in which controls and treatments are available, replication is affordable and easy to achieve, and good research design is possible. Field knowledge incorporates reality by investigating how environmental and ecological factors influence tools. Field knowledge is more difficult to attain because studies must be done at larger scales, for which appropriate controls are difficult to find; replication is expensive or logistically impossible and often not performed; and manipulation can be politically unfavorable.

For example, fladry (a rope with flags hanging at fixed intervals; Musiani and Visalberghi 2001) was proposed as a tool to exclude carnivores, in particular wolves, but little was know about fladry from both a methodological standpoint (Are wolves, bears, and other animals afraid of fladry? What is the optimal spacing of flags and height of rope?) and as a field tool (What factors influence the effectiveness of fladry? Does fladry work equally well in forested versus open habitats? How do native ungulate densities influence the effectiveness of fladry?).

To attain methodological knowledge, a number of studies were conducted in controlled settings to determine if fladry would exclude wolves (Musiani and Visalberghi 2001) and other carnivores (Shivik et al., 2003). The procurement of methodological knowledge about fladry was rapid and reliable: wolves are afraid of fladry and bears are not; optimal spacing of the flags is 25–50 cm; optimal rope height is 50 cm. However, these studies provided very little understanding about the application and effectiveness of fladry in the field.

To attain field knowledge, Musiani et al. (2003) carried out studies on two pastures in Alberta and one pasture in Idaho. To allow valid comparison between sites (external replication), these researchers made an effort to keep factors similar among sites. However, the experimental design was compromised to varying degrees at each site; for example, neither site had control or randomly allocated units. As a result, our understanding of fladry as a field tool is advanced but still nascent. We can say that fladry excluded wolves for about 60 days, but we have no real understanding of whether this is a general pattern or whether other influences, such as pack behavior or aspects of the environment affect the performance of fladry.

The majority of research published on tools used for carnivore-livestock conflict management is methodologically oriented, with much less published on field applications (see literature cited in Knowlton et al. 1999; Smith et al. 2000a, 2000b). Reasons for this discrepancy include issues of logistics (methodological studies are easier to perform); efficacy (many tools created in a lab simply do not work in the field, and thus field studies are not conducted); and a lack of understanding about necessary features of research design (i.e., control, replication, and randomization), causing abandonment of research protocol.

Because research is an inherently uncertain process and nonlethal tools are never 100% effective, there is generally a good chance that livestock will be killed by carnivores during a study. Combining this uncertainty with a strong intolerance of predators and misunderstanding of the importance of research design can lead to frustration and eventual abandonment of study design by producers. For example, a field experiment with a new scare device might compare depredation rates in unprotected (control) and protected (treatment) areas. If the treatment application were not 100% effective at repelling carnivores and livestock were killed, livestock producers could be strongly tempted to forget the study protocol and resort to other control techniques. Such intervention could cripple the study and lessen the ability to learn valuable information regarding scare devices—for example, it may be that scare devices reduce the killing of livestock but do not completely eliminate it.

Improving the Scientific Process

Difficulties associated with studying carnivores and livestock include working at large spatial scales, setting up field experimentation, attaining adequate replication, and maintaining a strong commitment to learning. Given these difficulties and given the importance of information for attaining a more harmonious balance between carnivores and livestock, what solutions may enable more rapid procurement of reliable information? Unfortunately there are no shortcuts to good experimental design. Manipulative experiments with adequate randomization, control, and replication offer the most effective means for gaining reliable information. But because reality dictates that optimal designs are rarely possible, researchers need to be opportunistic. They must take advantage of management actions to enhance study design and must maintain strong commitments to learning by developing collaborative and cooperative relationships with producers, managers, and other researchers.

The concept of adaptive management arose from the recognition that decisions regarding large-scale perturbations of ecosystems are often made without an understanding of how the ecosystem will respond to the perturbation. The central premise of adaptive management refers to a structured process of "learning by doing" that first integrates existing information into models, which are then used as guides in the design of management experiments (MacNab 1983; Walters and Holling 1990; Walters 1997). Properly employed adaptive management can take advantage of management decisions as manipulations that lead to better study design and more reliable knowledge.

Managers responsible for reducing conflict between carnivores and livestock are continually performing manipulations. Whether it is stringing fladry along a fence, placing out scare devices, selectively removing individuals, or altering husbandry, these actions can all be construed as manipulations. The key for effective learning is to take advantage of these perturbations by designing research programs around them and having the financial fortitude and personal commitment to follow through with monitoring. If possible, one should establish a control area to compare with treatment areas containing the manipulation, or if no separate control areas are available, it may be possible to monitor an area for a set period of time before the management action is implemented. Regardless of the design, the main premise of adaptive management is to link the actions that managers take on a routine basis with the ability of researchers to establish effective research designs and monitoring protocols.

Recognizing the potential to learn from management actions is a first step, but unless there is the will to make learning a priority, it is easy to ignore or abandon research. For example, radio-activated guards were designed as a nonlethal scare device for wolves (Shivik and Martin 2001). After some initial success indicating that scare devices may be effective deterrents for wolves (Breck et al. 2002), a number of groups purchased radio-activated guards and began using them for management purposes. With a little more effort devoted to planning and monitoring, many of these applications of radio-activated guards could have been used to more fully evaluate their effectiveness in a variety of environmental situations (a priority for nonlethal tools). As it was, most of the applications were never monitored consistently nor utilized in a way that facilitated learning.

Part of the problem is that many wildlife managers and livestock producers operate under a reactionary protocol that dictates immediate action, whereas researchers operate under a planning protocol that dictates slower planning and preparation. Thus managers are responding to problems such as a dead calf by utilizing tools to provide immediate assistance, whereas researchers may be slower to respond, to ensure that monitoring programs are properly established and that the study design is sound. This disconnect can be problematic unless there is a commitment to learning.

A commitment to learning can be defined as an innate curiosity combined with the fortitude to deal with the uncertainty of research. Regarding carnivore-livestock conflicts and the development of greater understanding and tools to manage these conflicts, it is assumed that most wildlife researchers and managers have a strong commitment to learning. It is critical that the producer whose resources are part of the research project share this commitment. Without such a commitment the research project is almost guaranteed to fail. Thus in situations in which a producer "just wants the problem solved," it may be best to apply existing techniques to resolve the problem and not spend resources trying to develop new information. When the producer is amenable to adaptive management and testing new methods, researchers should foster a meaningful and strong relationship through good communication and financial incentives. Finding these relationships may be one of the keys for meaningful advances.

Of the three cornerstones of experimental design (control, randomization, and replication), perhaps the most important is replication, both internal and external (Johnson 2002). Just as collaboration between researchers and managers will be key for utilizing management actions as manipulations in designed studies, so will collaboration between researchers from various organizations be important for increasing the ability to replicate studies, both internally and externally. To date, much of the research conducted on carnivore-livestock conflict resolution has been done by independent organizations or individuals with little interaction among one another. Where researchers are acting independently, it may be more difficult to incorporate adequate internal replication because of budgetary or logistical constraints. Having knowledge of the ongoing research programs of various researchers may enable opportunities to combine efforts and increase rigor.

For example, an article by Musiani et al. (2003) incorporates a number of separate studies of varying scale regarding fladry and its effectiveness for excluding wolves. A majority of the work was performed by researchers in Alberta, Canada, but a part of the article includes work conducted in Idaho by the author and others from USDA Wildlife Services. Before we conducted our research, we consulted with Dr. Musiani regarding logistical consideration of study design and technical aspects of fladry. The idea was to closely match the study design that the Canadian researchers had used so our results could act as a form of internal replication to their work. As it turned out, the results from Idaho closely matched the results from Canada and strengthened our understanding of fladry. If the two research groups had acted independently, the results from Canada would have been less conclusive and the results from Idaho might not have been published, because they lacked enough evidence to stand alone.

Collaboration among researchers will also be important for achieving external replication, in which entire studies are repeated. Johnson (2002) encouraged scientists to replicate studies using the same methods as in the original studies, to help eliminate variation due to methodology. Although scientific papers are written so studies can be repeated, it is usually helpful to communicate with the author who designed and published the study being used as a model. Subtle nuances may be gained regarding methodology and technique that help maintain consistency among studies.

Conclusion

Well-designed experiments are rare in the subdiscipline of carnivore-livestock conflict resolution because of the difficulties of manipulating systems, finding controls for treatment areas, and achieving replication. Identified needs to address these problems include more cooperation and collaboration between managers and researchers, long-term studies, and a commitment to the learning process. A possible mechanism to ensure that these needs are met might be the development of an international oversight committee that would (1) implement network strategies to prioritize research questions; (2) promote communication between scientists and managers; and (3) coordinate collaboration among participants.

Such a committee would encourage investigators at diverse institutions from around the world to communicate and coordinate their research efforts across organizational and geographical boundaries. Funding could be allocated to support the maintenance of long-term research sites, promote the development of students, and develop an Internet site that allows wildlife professionals to access the collective knowledge of the group. Thus managers and other researchers could rapidly determine the current state of knowledge regarding particular methods and technologies, ongoing studies, and recommended study designs that would enhance cooperation and collaboration. Whether such an oversight committee is possible remains to be seen, but researchers in the field of carnivore-livestock conflict resolution should nonetheless be aware of the benefits of collaboration and actively pursue such relationships.

Conflict between predators and livestock will exist as long as carnivore and livestock ranges overlap. For now there are no silver bullets to solving problems associated with predators and agriculture. Carnivores of all sizes are built to kill and eat other animals, and livestock are built to be eaten, having lost most of their antipredator instincts. It is intractable and untenable to imagine solving this problem so that carnivores do not kill livestock; however, it is reasonable to believe that we can optimize the interaction so that a minimum number of livestock are lost to predators and a minimum number of carnivores are lethally removed. Finding solutions will require well-planned and well-executed research to generate reliable knowledge regarding management solutions.

Literature Cited

Anderson, C. R. Jr., and F. G. Lindzey. 2003. "Estimating cougar predation rates from GPA location clusters," *Journal of Wildlife Management* 67:307–316.

Anderson, C. R. Jr., M. A. Ternent, and D. S. Moody. 2002. "Grizzly bear–cattle interactions on two grazing allotments in northwest Wyoming," *Ursus* 13:153–162.

Breck, S. W., R. Williamson, C. Niemeyer, and J. A. Shivik. 2002. "Non-lethal radio activated guard for deterring wolf depredation in Idaho: Summary and call for research," *Proceedings Vertebrate Pest Conference* 20:223–226.

Conner, M. M., M. M. Jaeger, T. J. Weller, and D. R. McCullough. 1998. "Impact of coyote removal on sheep depredation in northern California," *Journal of Wildlife Management* 62:690–699.

Cozza, K., R. Fico, and M. L. Battistini. 1996. "The damage-conservation interface illustrated by predation on domestic livestock in central Italy," *Biological Conservation* 78:329–336.

Fritts, S. H. 1982. *Wolf Depredation on Livestock in Minnesota.* Unites States Fish and Wildlife Service, Washington, DC. Resource Publication 145.

Johnson, D. H. 2002. "The importance of replication in wildlife research," *Journal of Wildlife Management* 66:919–932.

Knowlton, F. F., E. M. Gese, and M. M. Jaeger. 1999. "Coyote depredation control: An interface between biology and management," *Journal of Range Management* 52:398–412.

Landa, A., K. Gudvangen, J. E. Swenson, and E. Roskaft. 1999. "Factors associated with wolverine (*Gulo gulo*) predation on domestic sheep," *Journal of Applied Ecology* 36:963–973.

Linnell, J. D. C., J. Odden, M. E. Smith, R. Aanes, and J. E. Swenson. 1999. "Large carnivores that kill livestock: Do 'problem individuals' really exist?" *Wildlife Society Bulletin* 27:698–705.

MacNab, J. 1983. "Wildlife management as scientific experimentation," *Wildlife Society Bulletin* 11:397–401.

Mech, L. D., E. K. Harper, T. J. Meier, and W. J. Paul. 2000. "Assessing factors that may predispose Minnesota farms to wolf depredations on cattle," *Wildlife Society Bulletin* 28:623–629.

Musiani, M., C. Mamo, L. Boitani, C. Callaghan, C. C. Gates, L. Mattei, E. Visalberghi, S. Breck, and G. Volpi. 2003. "Wolf depredation trends and the use of fladry barriers to protect livestock in western North America," *Conservation Biology.* 17:1538–1547.

Musiani, M., and E. Visalberghi. 2001. "Effectiveness of fladry on wolves in captivity," *Wildlife Society Bulletin* 29:91–98.

Pyare, S., and J. Berger. 2003. "Beyond demography and delisting: Ecological recovery for Yellowstone's grizzly bears and wolves," *Biological Conservation* 113:63–73.

Rabinowitz, A. R. 1986. "Jaguar predation on domestic livestock in Belize," *Wildlife Society Bulletin* 14:170–174.

Robel, R. J., A. D. Dayton, F. R. Henderson, R. L. Meduna, and C. W. Spaeth. 1981. "Relationships between husbandry methods and sheep losses to canine predators," *Journal of Wildlife Management* 45:894–911.

Sacks, B. N., K. M. Blejwas, and M. M. Jaeger. 1999a. "Relative vulnerability of coyotes to removal methods on a northern California ranch," *Journal of Wildlife Management* 63:939–949.

Sacks, B. N., M. M. Jaeger, J. C. C. Neale, and D. R. McCullough. 1999b. "Territoriality and breeding status of coyotes relative to sheep predation," *Journal of Wildlife Management* 63:593–605.

Shivik, J. A., and D. J. Martin. 2001. "Aversive and disruptive stimulus applications for managing predation," *Proceedings of the Ninth Wildlife Damage Management Conference* 9:111–119.

Shivik, J. A., A. Treves, and P. Callahan. 2003. "Non-lethal techniques for managing predation: Primary and secondary," *Conservation Biology.* 17:1531–1537.

Smith, M. E., J. D. C. Linnell, J. Odden, and J. E. Swenson. 2000a. "Review of methods to reduce livestock depredation: I. Guardian animals," *Acta Agriculturae Scandinavica,* Section A—Animal Science 50:279–290.

Smith, M. E., J. D. C. Linnell, J. Odden, and J. E. Swenson. 2000b. "Review of methods to reduce livestock depredation: II. Aversive conditioning, deterrents, and repellents," *Acta Agriculturae Scandinavia,* Section A—Animal Science 50:304–315.

Stahl, P., J. M. Vandel, V. Herrenschmidt, and P. Migot. 2001. "Predation on livestock by an expanding reintroduced lynx population: Long-term trend and spatial variability," *Journal of Applied Ecology* 38:674–687.

Stahl, P., J. M. Vandel, S. Ruette, L. Coat, Y. Coat, and L. Balestra. 2002. "Factors affecting lynx predation on sheep in the French Jura," *Journal of Applied Ecology* 39:204–216.

Till, J. A., and F. F. Knowlton 1983. "Efficacy of denning in alleviating coyote depredations upon domestic sheep," *Journal of Wildlife Management* 47:1018–1025.

Walters, C. 1997. "Challenges in adaptive management of riparian and coastal ecosystems," *Conservation Ecology* 1(2):1–22. URL:http://www.consecol.org/vol1/iss2/art1.

Walters, C. J., and C. S. Holling. 1990. "Large-scale management experiments and learning by doing," *Ecology* 71:2060–2068.

CHAPTER 2

Characteristics of Wolf Packs in Wisconsin: Identification of Traits Influencing Depredation

Adrian P. Wydeven, Adrian Treves, Brian Brost, and Jane E. Wiedenhoeft

When carnivore and human activities intersect, one often sees economic losses or threats to human safety and recreation. Carnivores may be killed or removed as a result. Minimizing such conflicts could save resources, political goodwill, and rare or otherwise valuable carnivores. But effective reduction of depredations depends on anticipating the parties involved and the timing and location of conflicts. If one believes that all carnivores given an opportunity to prey on domestic animals will do so, then significant reduction in depredations may appear impossible, especially if conflicts are dispersed across broad regions and dense populations. However, the literature on human-carnivore conflicts tells a different story.

Not all carnivores with access to domestic animals will prey on them. Most individual large carnivores that range near livestock and humans do so without conflict for years (Tompa 1983; Polisar 2000; Stahl and Vandel 2001). Careful studies of radio-collared pumas (*Puma concolor*) and grizzly bears (*Ursus arctos*) suggest that some individuals avoid livestock, others remain nearby without attacking, and a subset preys on livestock (Jorgenson 1979; Suminski 1982; Bangs and Shivik 2001). Indeed, there is growing evidence that the timing of human-carnivore conflicts is nonrandom; that locations of conflicts share consistent characteristics; that the humans or domestic animals involved in conflicts share common features; and that the carnivores that cause problems are not a random subset of the population (Table 2.1). We find common patterns around the world, despite the involvement of many different carnivore taxa, varied husbandry systems, and culturally heterogeneous human populations. As a result, a great number of human-carnivore conflicts may be predictable.

Several caveats about Table 2.1 are warranted. We make no claim for the independence of each factor from others (e.g., unsupervised herds often wander into habitat providing cover for carnivores), nor do we argue that these relationships are always strong and pertinent to a particular predator-prey context. For example, many of the generalizations do not apply well to carnivore predation on humans. When carnivores specialize on humans, carnivores may hunt them by day, around settlements, and without regard to wild prey abundance (Corbett 1954; Turnbull-Kemp 1967; Brain 1981; Rajpurohit 1998). Another limitation of Table 2.1 is the omission of most information pertaining to which individual carnivores are more likely to be involved in conflicts with humans. As this is the subject of our case study, we treat the question separately in the next section.

Carnivores Involved in Conflicts

Although individual carnivores differ in their predisposition to conflict with humans, some predictable differences exist between ages, sexes, and social classes. The best evidence comes from coyotes (*Canis latrans*), where virtually all attacks on sheep have involved breeding pairs of coyotes (Knowlton et al. 1999). The predictability and management implications of this finding have been explored in detail previously (Sacks et al. 1999). On the other hand, relocated or dispersing bears (*Ursus* spp.) and lions (*Panthera leo*) are often involved in conflicts, perhaps more often than stable residents (Jorgensen et al. 1978; Fritts et al. 1985; Stander 1990; Linnell et al. 1997). It is not yet clear whether dispersers and transient carnivores are implicated in more conflicts because they are more easily captured (Sacks et al. 1999), while the actual culprits escape, or if real differences exist between taxa in the involvement of residents and transients.

Many authors have argued that infirm or injured carnivores are more often involved in conflicts—for example, jaguars (*P. onca;* Rabinowitz 1986; Hoogesteijn et al. 1993); Indian leopards (*P. pardus;* Corbett 1954); and tigers (*P. tigris;* Corbett 1954). However, the evidence remains equivocal (Aune 1991; Faraizl and Stiver 1996; Linnell et al. 1999; Treves and Naughton-Treves 1999; Treves et al. 2002). An alternative explanation is that carnivores already predisposed to range near humans or settlements

TABLE 2.1
Common factors that increase the risk of carnivore predation on domestic animals

Variation in Risk and Predictability of Conflicts		
Timing of Conflicts	Lower Risk	Higher Risk
Relative to guards	Active supervision	Unsupervised
Circadial	Daylight	Darkness
Seasonal	Few, well-defended small domestic animals	Many vulnerable small domestic animals
	Domestic animals confined	Unconfined
Location of Conflicts	Lower Risk	Higher Risk
Habitat	Abundant or vulnerable wild prey	Scarce or well-defended wild prey
	Open, no concealment for carnivores	Forested, closed, or rough terrain
	Close to development (settlements, roads, lights)	Far from developed areas
Domestic animal activities	Circumscribed	Free-roaming, stray
	Far from wild prey	Near wild prey
	Far from garbage and carcasses	Near garbage and carcasses
	Far from center of carnivore territories	Near carnivore dens
	Far from protected areas	Around protected areas
Participants in Conflicts	Lower Risk	Higher Risk
Humans	Vigilant, nearby	Unwary, distant
	Best husbandry	Negligent husbandry
Domestic animals	Well-guarded	Unguarded
	Adult, large	Young, small
	Healthy, strong	Infirm or pregnant
	Smaller herds	Larger herds
Carnivores	Wary of humans or naive of human foods	Habituated to or previously fed human foods
	Healthy, prime-age carnivores	Rabid, infirm, young or aged

Sources: Jorgensen et al. 1978; Jorgensen 1979; Robel et al. 1981; Bjorge and Gunson 1983; Mech et al. 1988; Fritts and Paul 1989; Aune 1991; Fritts et al. 1992; Quigley and Crawshaw 1992; Hoogesteijn et al. 1993; Jackson et al. 1996; Meriggi et al. 1996; Ciucci and Boitani 1998; Kaczensky 1999; Landa et al. 1999; Mech et al. 2000; Polisar 2000; Rajpurohit and Krausman 2000; Oakleaf et al. 2003; Ogada et al. 2003; Treves et al. 2004.

are also more likely to be those injured by traps, vehicles, and gunshots, producing a spurious correlation between injury and conflict with humans.

Many of the findings above and their exceptions suggest that attributes of carnivores that predispose them to conflict are often taxon-specific or even specific to individuals. For example, Linnell and colleagues (1999) noted a tendency for male carnivores to be involved in more conflicts than females, although they stressed that this tendency did not apply to wolves (*Canis lupus*). Here we describe a case study from the state of Wisconsin that illustrates how managers and researchers might use predisposing factors to predict and avert wolf attacks on domestic animals (hereafter referred to as depredation).

Background

In the late 1970s, wolves recolonized Wisconsin from Minnesota, their last significant refuge in the contiguous United States (Young and Goldman 1944; Thiel 1993; Wydeven et al. 1995). Wisconsin's first confirmed wolf depredation happened in 1976, but depredations remained rare until the 1990s (Treves et al. 2002). Since 1990, the number of wolves and the number of depredations in Wisconsin have risen steadily to a minimum estimate for late winter 2002 of 323 wolves, with a cumulative total of 572 domestic animal losses in 126 verified incidents (Treves et al. 2002; Wydeven et al. 2002).

Wolf depredation fits three functional categories in Wisconsin. Most common, wolves entered fenced pastures or poultry areas to prey on livestock (Treves et al. 2002). Wisconsin does not have free-ranging livestock herds or unfenced grazing allotments as in other areas of the country. Second most common, hunting dogs were killed on public land when they roamed into wolf pack rendezvous areas or denning sites. The hunting dogs, often valued at $2,000–$5,000, were typically monitored remotely (radio collar or other means) by their owners as they roamed for kilometers. The third and rarest form of depredation occurred when one or more wolves entered a fenced, forested area containing farm deer (*Odocoileus virginianus*) intended for trophy hunters (Treves et al. 2002).

The Wisconsin Department of Natural Resources (DNR) has inferred which pack or individual is responsible for depredation, but individual wolves causing depredation are rarely identified because only about 20% of the wolf population has been radio collared at any one time (Wisconsin

DNR 1999). Nevertheless, these inferences are robust because the Wisconsin wolf population has been intensively monitored by radiotelemetry, winter track surveys, and summer howl surveys ever since its return to the state (Wydeven et al. 1995). Thus, verified wolf attacks can often be attributed to a known wolf pack based on location, past history of the pack, and other contextual information (Wydeven and Wiedenhoeft 2000; Wydeven et al. 1995, 2002). The criteria used by field verifiers and cooperating agencies have been detailed previously (Willging and Wydeven 1997; Treves et al. 2002). In rare cases, conflicts may be incorrectly attributed to nearby wolf packs when other carnivores are actually involved (free-roaming dogs [*Canis familiaris*], wolf-dog hybrids, transient wolves, black bears [*Ursus americanus*], coyotes, etc.). Mitigating this, wolves may cause some disappearances of domestic animals without evidence (Oakleaf et al. 2003). Based on direct confirmation of hybrids and transient wolves (confirmed wolf attacks arising farther than 5 km from any known pack), we estimated that <10% of verified wolf depredations actually involve other carnivores.

In previous analyses of verified depredations across Minnesota and Wisconsin, researchers found that farms with larger landholdings and greater numbers of cattle faced higher risk of wolf depredation than their unaffected neighbors with similar operations (e.g., both producing beef cattle) (Mech et al. 2000; Treves et al. 2004). We also found broader landscape predictors of past wolf attacks on livestock; namely, affected townships (square survey blocks of 92.16 km^2) contained more pasture, more deer, fewer crops, and fewer roads than their unaffected neighbors (Treves et al. 2004). Building on this work, we predicted that problem packs occupy territories with landscape features that promote encounters with dogs (hunting dogs or domestic dogs) or livestock (bovids, equids, ovids, poultry, and farm deer). We also drew on the literature from coyotes to predict that demographic features of wolf packs distinguish problem from non-problem wolf packs (Knowlton et al. 1999; Sacks et al. 1999). Namely, we examined whether pack size or pup production predicts which packs will be blamed for predation on domestic animals. We also explored the relationship between depredations and a wolf pack's tenure in its home range.

Methods

Wolves were live-trapped and radio collared following established procedures (Mech 1974; Wydeven et al. 1995). Only wolves weighing >13.6 kg were fitted with radio collars. Radio-collared wolves were generally located once per week from the air by DNR pilots using fixed-wing aircraft, but dispersing and recently translocated wolves were sometimes located 2–3 times per week. Trapping for population monitoring was conducted in late April–September from 1979 to 2003. Additionally, USDA Wildlife Services live-trapped wolves at sites of verified depredation (February–October from 1991 to 2003); these were fitted with radio collars and translocated across northern Wisconsin. Between 1991 and 2003, translocated wolves caused subsequent confirmed livestock depredation once in >30 translocations (Wisconsin DNR, unpublished data).

Late-winter pack size (before pups are born) was estimated annually from 1979. About half of the estimates of pack size were collected by DNR pilots' visual observations of radio-collared individuals and their associates. When these data were not available, the DNR used winter track surveys to estimate pack size. Winter track surveys sometimes also provided evidence of breeding (double raised-leg urinations or blood left in urine marks) (Rothman and Mech 1979), or the presence of two adults defending a territory (scats, scratching, and raised-leg urinations) (Peters and Mech 1975).

Since 1995, DNR biologists have supplemented their own winter track counts and surveys with data provided by 55 to 135 volunteers. DNR biologists verified observations made by volunteers before pack counts were confirmed. In addition, summertime howl surveys provided information on pup presence or absence, location of rendezvous sites, location of non-collared packs, and the rise of new packs (Harrington and Mech 1982). Howl surveys helped locate wolves and determine the presence of pups but were considered unreliable for accurate counts of wolves beyond 2 adults or 2 pups (Harrington and Mech 1982).

Additional data were collected during trapping operations when the DNR biologists were able to observe pack members directly. The pup count used here is therefore an estimate based on a combination of direct and indirect evidence collected in both the summer and winter. As a result, pup count is statistically related to total pack size because DNR biologists estimated past pup production from current- and previous-year counts of adults and yearlings. Pup count estimates for packs in winter were often a range of values (e.g., 4–6); we used the median value to estimate total pups per individual pack.

We used winter pack size, pup counts, and wolf pack tenure (length of residence in an area) as our demographic indices to test whether wolves involved in predation on domestic animals could be discriminated from others. We used the average pack size and average pup count for each pack, calculated over all years that pack was monitored, rather than the value at the time of depredation, for two reasons. Most packs involved in a verified depredation were implicated in >1 incident, so we chose the average of our demographic indices rather than focusing on a single year's value. Secondly, annual pup count and pack size estimates reflected 2 or more time points, each of which had potential error. Had we focused on a single year for our demographic parameters, we might have increased the error, whereas by taking the average of several years we derived a more robust estimate of the pack's central tendency during the years in question.

Home range area for the period April 15–September 14 from 1999 to 2003 was determined with the minimum convex polygon method (Mohr 1947). Isolated radio locations over 5 km from other points were considered extraterritorial moves (Fuller 1989). When two separate clusters of radio locations existed with regular travel between them, then areas in between were considered part of the home range, regardless of distance, as long as both clusters did not occur in another pack territory. Home range areas were calculated only for wolves that occupied stable ranges for 1 year or more, and did not include wolves that dispersed. For packs without radio-collared animals, DNR biologists superimposed the population average home range on noncollared pack locations recorded in winter track surveys (Wydeven et al. 1995, 2002). This procedure might generate some random error in our analyses of landscape features for those packs without radio-collared individuals. However, we have employed a 5 km buffer around the estimated and known home range for each pack, in order to encompass error in these estimates and to account for occasional extraterritorial movements of individual wolves. Hereafter, "pack area" refers to the calculated or estimated home range plus a 5 km buffer.

Analysis

Each pack with landscape data contributed one sample to each analysis for a total sample of 80. Wisconsin has had more packs than this, but some were known for only 1 year before they were removed by control operations or disappeared. Our response variables were nominal (e.g., involvement in dog depredation); continuous (the number of incidents); or categorical (the type of animal preyed upon—scored as none, dog, livestock, or both). We used nonparametric analyses for univariate tests (Mann-Whitney U, Kruskal-Wallis H, Wilcoxon signed-ranks, and Spearman rank correlation analysis rho) but relied on parametric regression techniques for multivariate tests. Significance was set at $p = 0.05$.

We used GIS (Geographic Information System) to analyze landscape features within wolf pack areas, using the USGS 1992/1993 land cover classification of the entire United States (Vogelmann et al. 2001). Some land cover classes were pooled: unusable = all residential land cover classes, bare rock, barren, and urban grassy areas; crops = row crops and small grains. Intercorrelated land cover classes were examined individually or summed (e.g., forested wetland + emergent wetland). Percentages for land cover were transformed using the arcsine–square root transformation (Sokal and Rohlf 1981); road density was estimated in km/km^2 (TIGER/Line files 1992) using methods described previously (Mladenoff et al. 1995).

Results

Wolf pack summer home ranges averaged 79.4 km^2 based on radio-collared adults with >19 radio locations. The home ranges and 5 km buffers varied in landscape features (Table 2.2). All wolf packs had some pasture, hayfields, or crops within the area encompassed by their territory, plus a 5 km buffer around it, indicating that all wolf packs could potentially move onto agricultural lands. Hunting coyotes with dogs is permitted throughout the state most of the year, so all wolf packs might encounter free-roaming dogs.

From 1976 to 2002, 31 of the 80 (38.8%) wolf packs included in our study were implicated in 82 incidents of depredation (11 packs on livestock only, 10 on dogs only, and 10 on both types of domestic animals). The number of independent incidents is only an estimate because some cases of depredation may have involved repeated entries and departures that were subsequently pooled into a single report by the DNR. Notwithstanding, individual packs were implicated in 0–8.5 incidents each (fractions were assigned when either of two packs might have been responsible for an incident). Ten of 31 (32%) packs were implicated in only 1 incident, while 21 (68%) were blamed for >1 incident. The number of incidents was positively correlated to tenure (the number of years each wolf pack was confirmed

TABLE 2.2
Landscape features of 80 wolf pack areas in Wisconsin as a percentage of home range area plus a 5 km buffer

Feature	Average		±1 Std. Dev.	Range
Road density	0.54	km/km^2	0.18 km/km^2	0.18–1.19 km/km^2
Open water	4.3	% of area	4.9%	0.2–29 %
Unusable	0.6	% of area	3.4%	0.0–30 %
Transitional	1.4	% of area	5.0%	0.0–43 %
Deciduous	49.9	% of area	14.0%	5.2–76 %
Evergreen	6.9	% of area	5.3%	0.4–32 %
Mixed forest	10.5	% of area	4.1%	0.0–8 %
Shrub	0.0	% of area	0.1%	0.0–1 %
Grass	0.5	% of area	0.8%	0.0–5 %
Pasture/hay	3.1	% of area	2.9%	0.2–14 %
Crops	3.7	% of area	4.1%	0.4–32 %
Woody wetlands	14.5	% of area	10.6%	0.4–41 %
Emergent wetlands	4.6	% of area	4.0%	0.0–23 %

present in its territory). This relationship hints that any wolf pack may cause a depredation eventually, or that wolf packs undergo internal changes over time that lead to depredations. Of 33 wolf packs studied for >5 years, 15 (45%) were implicated in depredation. Hence, many wolf packs did not prey on domestic animals despite having access for several years.

Until the mid-1990s, wolf depredation had been relatively uncommon in Wisconsin, but depredations on livestock occurred every year from 1995 to 2002, and depredation on dogs occurred every year from 1996 to 2002. Between 1995 and 2002, a mean of 7% (±3%) of packs in the state were involved in depredation on livestock and a mean of 10% (±5%) of packs were implicated in depredation on dogs. From 1997 through 2002, 3% (±1%) of packs were implicated in depredations on both livestock and dogs.

Restricting our analysis to problem packs, the annual rate of incidents (a pack's number of incidents divided by pack tenure) averaged 0.43 incidents per year (±5%)0.10, n = 21, range 0.08–2.0) for livestock depredation, while the average rate for depredation on dogs was 0.61 incidents per year (±0.10, n = 21, range 0.08–3.0). In other words, packs implicated in a dog depredation repeated this in 45–76% of subsequent years, whereas packs implicated in a livestock depredation repeated less often (33–53% per year). Although the average annual rate of depredation on dogs did not differ from that of livestock depredation (Mann-Whitney U test Z = 0.23, p = 0.83), the rate of dog depredation was more variable (test of homogeneity of variance: F = 0.44, p = 0.030). One pack (Shoberg Lake) killed dogs 5 years in a row, and another pack (Kidrick Swamp) killed dogs 4 years in a row. The longest series of livestock depredations was 3 years (Chase Brook Pack), but packs that caused livestock depredation were subject to control live-trapping and translocation.

Average pack size ranged from 2 to 6 adults and yearlings per pack (n = 80 packs with tenure of 1–12 years). The pack sizes reported here are typical for wolf packs that prey mainly on white-tailed deer (Mech 1970). Tenure correlated positively with average pack size (rho = 0.33, p = 0.0033) and average pup count (rho = 0.48, p = <0.0001), reflecting how dispersers of both sexes met and formed new packs, then retained yearlings as helpers after the pair bred successfully (Mech 1970). Average estimates of pup numbers in winter ranged from 0 to 3.8 for individual packs (mean 1.4 ±1.0, n = 80 packs with breeding tenures of 1–12 years). There was no consistent difference between pup counts in the year with a depredation and that for the previous year (considering problem packs only, Wilcoxon signed-ranks p = >0.05 for all types of depredation). Also, there was no difference between average pack size and pack size in the year of a depredation (p = >0.05 for all types of depredation).

With univariate tests, we examined our three demographic variables (average pack size, average pup count, and tenure) to see if these discriminated problem packs from others or predicted type and intensity of conflict (Table 2.3). Average pack size discriminated problem packs that preyed on dogs from all other packs, and differentiated the four different types of depredation history (dog, livestock, both, none). Tenure was the only useful demographic variable for predicting the total number of incidents.

We combined the landscape features in and around wolf pack areas (see Table 2.2) with the strongest demographic variables from Table 2.3. Because of the large number of potential predictors, we performed the regressions in two steps: first, we included all predictors; then, in step two, we dropped those predictors with partial t-values from 1.5 to −1.5. The logistic regression model discriminating wolf packs implicated in dog depredation from other wolf packs was strongest, explaining 24% of the variation; the other models were significant but explained only 12–16% of variation (Table 2.4).

Consistent with our univariate tests, wolf packs implicated in depredation on dogs and those implicated in a greater number of such

TABLE 2.3
Associations between wolf pack attributes and involvement in depredations

Demographic characteristics	Problem Packs That Attacked			Number of Dog Incidents	Number of Livestock Incidents	Total Number of Incidents
	Dogs vs. Others	Livestock vs. Others	Dogs, Livestock Both, or None			
Average pack size	4.1 vs. 3.3 $p = 0.00061^1$	3.4 vs. 3.5 NS^1	4.5, 3.1, 3.6, 3.3 $p = 0.0020^2$	rho = 0.37 $p = 0.0009^3$	rho = -0.04 NS^3	rho = 0.23 $p = 0.039^3$
Average pup survival	2.0 vs. 1.3 $p = 0.0018^1$	1.5 vs. 1.4 NS^1	2.4, 1.4, 1.7, 1.2 $p = 0.0074^2$	rho = 0.34 $p = 0.0026^2$	rho = 0.07 NS^3	rho = 0.28 $p = 0.014^3$
Pack tenure (years)	6.5 vs. 4.6 NS^1	6.1 vs. 4.7 NS^1	5.9, 5.3, 7.0, 4.5 NS^2	rho = 0.20 NS^3	rho = 0.20 NS^3	rho = 0.23 $p = 0.038^3$

Statistical tests used:
[1] Mann-Whitney U-test, corrected for ties.
[2] Kruskal-Wallis, df = 3.
[3] Spearman Rank; NS = not significant.

incidents were larger than other packs (Table 2.4). Tenure remained in the livestock models (t = >1.5), but not significantly—this may reflect its association with forested habitats because the long-studied wolf packs were the first to recolonize the northwestern, forested portions of the state (Table 2.4).

Among landscape predictors, the proportion of deciduous forest, evergreen forest, and transitional vegetation remained in the predictive models. In particular, the areas used by packs implicated in dog depredation had more evergreen forest than those of other packs, whereas those implicated in livestock depredation had more deciduous forest than other packs. These distinctions between forest types may not be meaningful, given that

TABLE 2.4

Wolf pack demography and range features associated with depredations. Only those predictors with t-values <1.5 or >-1.5 were retained in the models.

Problem Packs That Attacked Dogs vs. Other Wolf Packs		
Predictors	t	p
Average pack size	3.78	0.0003
Evergreen forest (% by area)	2.39	0.019
Open water (% by area)	−1.80	0.076
Deer density	−1.77	0.081
Using logistic regression (r = 0.49, n = 80, p = 0.0012)		
Number of Dog Predation Incidents		
Predictors	t	p
Average pack size	2.41	0.018
Transitional vegetation (% by area)	1.93	0.058
Using multiple regression (r = 0.34, n = 80, p = 0.008)		
Problem Packs That Attacked Livestock vs. Other Wolf Packs		
Predictors	t	p
Deciduous forest (% by area)	2.84	0.0057
Tenure	1.63	0.11
Using logistic regression (r = 0.35, n = 80, p = 0.0065)		
Number of Livestock Predation Incidents		
Predictors	t	p
Deciduous forest (% by area)	2.83	0.006
Area of territory + 5 km	−2.18	0.033
Transitional vegetation (% by area)	1.90	0.061
Tenure	1.51	0.14
Using multiple regression (r = 0.40, n = 80, p = 0.0093)		

the satellite imagery dates to 1992–1993 (Vogelmann et al. 2001) and that forest composition changes over time with human management and natural succession.

The only other significant landscape predictor was the size of the wolf pack area, which was negatively associated with the number of incidents of livestock depredation (Table 2.4). This suggests that wolf packs with smaller areas might encounter livestock more often or might have less access to vulnerable wild prey and thus select alternate prey like livestock, or that use of livestock permits a smaller home range.

We return to the observed differences in average pack size to consider the predictability of depredation. The type of domestic animal depredation was associated with average wolf pack size (Table 2.3; Figure 2.1). This pattern reflects that packs involved in depredations on dogs were larger (4.5, n = 10) than (a) those packs never implicated in depredations (3.3, n = 49); (b) those packs that attacked livestock only (3.1, n = 11); and (c) those packs that attacked both livestock and dogs (3.6, n = 10). Only one pack with an average size below 3 (North Empire) was blamed for dog depredation, and this wolf pack arose from the fission of a larger pack. No Wisconsin wolf pack was implicated in livestock depredation when the average pack size was below 2.2 or above 4.8 (see Figure 2.1). Capture and translocation programs have removed animals from chronic livestock depredation sites (Treves et al. 2002). At one Wisconsin farm, 22 wolves were removed from at least 3 packs over 4 years (DNR unpublished data). Presumably, the average pack size decreased each time a translocation occurred.

Although the average size of packs involved in depredations on livestock or both livestock and dogs was not statistically distinguishable from the average size of nonproblem packs, both categories of problem packs were less variable in size (F = 0.28, p = 0.013, and F = 0.33, p = 0.048, respectively). Examining only problem packs, the number of incidents of depredation on dogs increased with wolf pack size (Spearman rho = 0.49, p = 0.010), whereas the number of incidents of depredation on livestock decreased as pack size increased (rho = –0.63, p = <0.0001).

If we assume that any wolf pack may attack a dog if given the opportunity, we can use the relationship with pack size to forecast risk by wolf pack. In Figure 2.2, we graph the annual risk of wolf attack on dogs according to wolf pack size, assuming that the other significant predictor (proportion of forest; Table 2.4) remained constant.

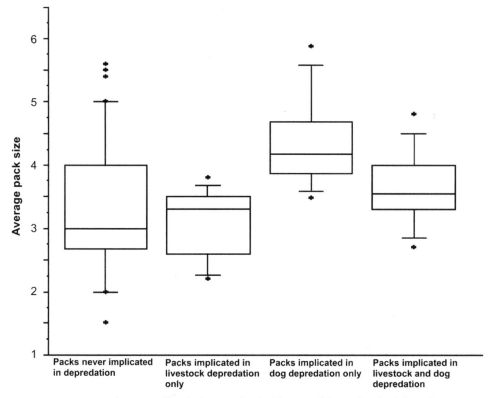

FIGURE 2.1 Average wolf pack size associated with types of domestic animal depredation. The points (✦) depict data points lying beyond the 95th percentile. The upper bars span the 75th to 95th percentiles. The lower bars span the 5th to 25th percentiles. The boxes span the first to third quartile. The median is shown by the horizontal line within each box.

Discussion

Most wolf packs were never implicated in depredation on domestic animals, although all of them had some access to dogs or livestock. In Wisconsin, wolf depredations take two very different forms: (1) wolves coming onto fenced areas on private land, killing livestock, poultry, or farm deer; and (2) wolves killing hounds on public lands. In the first situation, the wolves appeared to have been seeking food, judging from consumption of the carcasses. In the second situation, wolves probably reacted to dogs as trespassers in territorial defense, or as competitors. This type of

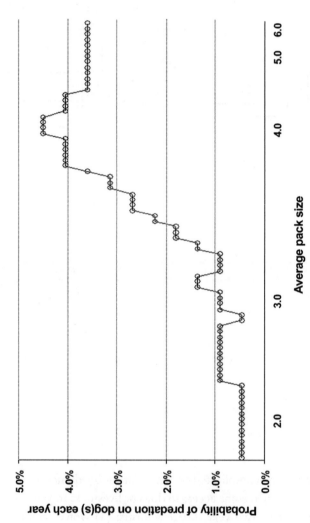

FIGURE 2.2 The annual probability of wolf pack predation on dogs as a function of average wolf pack size. For each point on the x-axis, we examined the depredation history for the 20 wolf packs closest in size to this point and computed the proportion of those wolf packs ever implicated in dog depredation. This corresponds to a moving average. To calculate probabilities on the y-axis, we multiplied the proportions above by the average annual rate of depredation on dogs (total incidents divided by total wolf pack years observed in Wisconsin).

depredation does not occur in adjacent Minnesota, where hunting of bear with hounds is not allowed and most wolf depredation on dogs is near homes (Fritts and Paul 1989). We believe most of Wisconsin's dog depredations occur when dogs get too close to wolf pups at summer rendezvous sites, a situation in which wolves would likely be aggressive to other large carnivores (Murie 1944). In Denali National Park, aggressive encounters with neighboring wolf packs is the primary cause of wolf mortality (Mech et al. 1998), and in areas where wolves and coyotes are sympatric, wolves cause substantial coyote mortality (Paquet 1991; Arjo and Pletscher 1999).

Because dog and livestock depredations appear different behaviorally, we anticipated differences in predictability and in the characteristics of the wolf packs involved. Although ~80% of wolf packs occupy areas open to bear hunting with hounds in northern Wisconsin, and all of the state is open to hunting of coyotes with hounds, only 4–10% of wolf packs in the state are implicated in dog depredation annually. Larger packs, with more pups, were more likely to attack dogs, while smaller packs with smaller home ranges were more often implicated in livestock depredation. The wolves implicated in more incidents of depredation on dogs had more conifer cover, whereas packs depredating more on livestock had more deciduous forest cover. Research on wolf territory establishment in Wisconsin revealed that areas selected by packs had more conifer cover and less deciduous forest (Mladenoff et al. 1995). Also, deciduous forests tend to occur on better soils and are thus more likely associated with agricultural land. In sum, the wolf packs that caused problems were distinct from others in the population, both demographically and by characteristics of their home ranges.

Although our analysis provides some predictability in determining likely wolf depredation (see Figure 2.2), the power is relatively low, and the probability that any one pack will cause depredation on dogs is less than 5% annually. The occurrence of prior depredations appears to be a better predictor, because wolves that caused depredation on dogs repeated this in 45–76% of succeeding years. Livestock depredations were repeated by the same pack in 33–53% of subsequent years. Wolf depredation on livestock may be more predictable when farm and landscape attributes are considered (Treves et al. 2004).

When wolves recolonized Wisconsin, they initially settled forested, remote areas of the state, with very little farm land (Mladenoff et al. 1995), and depredations remained rare before the mid-1990s. As wolves continued to expand across the state, packs began to occupy areas with more farmland (Mladenoff et al. 1997, 1999). Possibly some threshold level of encounters is necessary before wolves "switch" (Murdock 1969) from wild prey to livestock. If this is true, increasing encounter rates may increase the likelihood of prey switching.

Our findings are reminiscent of studies of coyotes, in which breeding, territorial pairs were more likely to be involved in conflicts (Knowlton et al. 1999; Sacks et al. 1999). However, we found that average wolf pack size was a stronger predictor of involvement in conflicts than our estimate of pup survival. This may reflect winter pup counts and pack estimates, while depredation to provision pups would likely occur in spring and summer. Without better estimates of pup numbers in late spring and summer, we find it difficult to evaluate whether livestock depredations reflect the need to feed pups during times of wild prey scarcity. The association between wolf pack size and depredation on hunting dogs is unlikely to reflect the greater nutritional demands of larger wolf packs because dog carcasses were rarely fed upon; rather, it may reflect that larger wolf packs defend territories more vigorously. A search for causal explanations must await more detailed behavioral studies, but in the meantime, we believe our findings have value for managers and stakeholders.

If large carnivores and humans are to coexist with minimal conflict, we will need to arm carnivore managers and other stakeholders with tools to help them predict risk, reduce conflicts, and manage the aftermath effectively. Our case study estimates the risk of depredation faced by livestock producers and hunters using dogs when they operate near wolf packs. Information such as this can be valuable for several purposes.

Because unpredictability inflates perceived threats, any improvement in predictive ability makes conflicts appear more tractable. The DNR could provide hunters with general maps and information on wolf packs, including past depredation histories. Maps need not show precise locations of wolf den sites or rendezvous areas or information on radio locations, but they should provide enough detail that hunters can avoid such areas if they choose. Information should also be provided to help hunters identify rendezvous sites and wolf sign in the field.

Information on wolf pack involvement in depredation can also aid managers. The DNR now uses lethal control to manage problem wolves on private land (Wisconsin DNR 2002). Such control actions will only be done in response to depredations on livestock and pets on private land. Control operations will not be conducted following depredations on dogs within public land (Wisconsin DNR 1999, 2002). Because the current guidelines resemble those used in neighboring Minnesota (Minnesota DNR 2001), we can estimate the impact of Wisconsin's lethal control program. In recent years, 4–9% of the Minnesota wolf population has been removed because of control action (Mech 1998), but wolves were captured successfully in only 53% of sites where trapping was attempted. Under current conditions in Wisconsin, trapping is expected to affect <7% of Wisconsin wolf packs, and only after >1 confirmed depredation (Wisconsin DNR 2002). Thus in most years, only a very small proportion of the population will be removed through control trapping—much less than the 28–30% sustainable harvests deemed possible for wolf populations (Fuller 1989).

Between 1997 and 2003, the DNR paid out $241,230 in compensation for wolf predation, of which ca. $28,675 was paid for livestock annually and ca. $15,030 for dogs annually (R. Jurewicz, pers. comm.). This does not include the costs of live-trapping and translocation operations at chronic livestock depredation sites. Several alternatives to compensation and reactive control could be explored within these cost brackets, and our data on wolf pack involvement in depredation would help to focus such experiments.

Preventive technologies offer an alternative to lethal control and compensation. The merits of guarding, various deterrent devices, and fencing have all been examined at one time or another (Treves and Karanth 2003), but their application to wolves remains limited or little studied to date (Coppinger et al. 1988; Andelt 2001; Bangs and Shivik 2001; Musiani et al. 2003; Shivik et al. 2003). As a result, the cost-effectiveness of prevention is hard to estimate at present. Nevertheless, preventive methods should be tested at high-risk sites or among chronically depredating packs, to permit comparison with existing compensation and control programs (Treves et al. 2002).

For example, DNR managers are contemplating nonlethal methods for controlling problem wolves. Canid shock collars with remote triggering devices have been used in a few cases to keep wolves out of specific areas (Andelt et al. 1999; Schultz et al., unpublished data). Sterilization techniques (Haight and Mech 1997; Knowlton et al. 1999; Bromley and Gese 2001) may be used proactively against wolf packs living where most livestock depredations are likely to occur, or for areas with chronic depredation problems in the past. The sterilization could be temporary and relaxed periodically to maintain pack stability. Both methods might allow packs to maintain their positions and keep other wolves out, while reducing the incentives (i.e., feeding pups) to approach livestock-producing properties.

Finally, information on the predictability and location of depredating wolf packs may help managers to designate zones under which different management techniques might be applied. Such information could be incorporated into plans for wolf harvests or adaptive management aimed at protecting source wolf packs (*source packs*—packs originating in areas where the rates of population reproduction exceed both mortality and carrying capacity, so that the wolves immigrate to new areas; Pulliam 1988) and discriminating against sink wolf packs (*sink packs*—packs in areas in which losses from mortality or emigration exceed the levels of reproduction or immigration required to increase or stabilize a population; Pulliam 1988).

Conclusion

We have presented a case study of wolf depredation on domestic animals that increases the predictability of such conflicts and thereby opens new management options. Our demographic estimates of risk from a given wolf pack can readily be combined with preexisting locational and temporal predictors of conflict to focus outreach and reduce depredation problems. Also, our information on characteristics of packs that attack dogs should be useful to hunters who use hounds. Minimizing depredations is essential to maintaining public goodwill and conserving resources and valued wildlife. Our case study was possible only because Wisconsin invested substantially in the monitoring of its wolf population and the investigation of depredation claims. Similar analyses might be profitably done on other group-living carnivores where demographic features are suspected to influence depredation behavior.

Acknowledgments

Wolf population monitoring was supported by endangered species funds from the United States Fish and Wildlife Service, Federal Aid in Wildlife Conservation, the Chequamegon-Nicolet National Forest, the Wisconsin DNR Endangered Resources Fund, the Wisconsin Department of Transportation, the Timber Wolf Alliance, the Timber Wolf Information Network, Defenders of Wildlife, and private donations. B. E. Kohn,

R. N. Schultz, R. P. Thiel, and other Wisconsin DNR personnel helped monitor the Wisconsin wolf population, as did graduate students at University of Wisconsin—Stevens Point. We also are grateful to K. Thiel of USDA Wildlife Services, who provided essential support in organizing the depredation database. L. Naughton, R. Jurewicz, R. Willging, and R. Rose provided assistance in diverse ways. A. Treves was supported by the Center for Applied Biodiversity Science of Conservation International and by Environmental Defense during writing and analysis. Brian Brost was supported by a University of Wisconsin Hilldale Undergraduate Research Grant.

Literature Cited

Andelt, W. F. 2001. "Effectiveness of livestock guarding animals for reducing predation on livestock," *Endangered Species Update* 18:182–185.
Andelt, W. F., R. L. Phillips, K. S. Gruver, and J. W. Guthrie. 1999. "Coyote depredation on domestic sheep deterred with electronic dog-training collar," *Wildlife Society Bulletin* 27:12–18.
Arjo, W. M., and D. H. Pletscher. 1999. "Behavioral responses of coyotes to wolf recolonization in northwestern Montana," *Canadian Journal of Zoology* 77:1919–1927.
Aune, K. E. 1991. "Increasing mountain lion populations and human-mountain lion interactions in Montana," in *Mountain Lion–Human Interaction Symposium and Workshop*, ed. C. E. Braun, 86–94. Colorado Division of Wildlife, Denver, CO.
Bangs, E., and J. Shivik. 2001. "Managing wolf conflict with livestock in the northwestern United States," *Carnivore Damage Prevention News* 3:2–5.
Bjorge, R. R., and J. R. Gunson. 1983. "Wolf predation of cattle on the Simonette River pastures in northwestern Alberta," in *Wolves in Canada and Alaska: Their Status, Biology and Management*, ed. L. N. Carbyn, 106–111. Canadian Wildlife Service, Edmonton, AB.
Brain, C. 1981. *The Hunters or the Hunted? An Introduction to African Cave Taphonomy.* University of Chicago Press, Chicago.
Bromley, C., and E. M. Gese. 2001. "Effect of sterilization on territory fidelity and maintenance, pair bonds, and survival rates of free-ranging coyotes," *Canadian Journal of Zoology* 79:386–392.
Ciucci, P., and L. Boitani. 1998. "Wolf and dog depredation on livestock in central Italy," *Wildlife Society Bulletin* 26:504–514.
Coppinger, R., L. Coppinger, G. Langeloh, L. Gettler, and J. Lorenz. 1988. "A decade of use of livestock guarding dogs," *Proceedings of the Vertebrate Pest Conference* 13:209–214.
Corbett, J. 1954. *The Man-Eating Leopard of Rudraprayag.* Oxford University Press, London.
Faraizl, S. D., and S. J. Stiver. 1996. "A profile of depredating mountain lions," *Proceedings of the Vertebrate Pest Conference* 17:88–90.
Fritts, S. H., and W. J. Paul. 1989. "Interactions of wolves and dogs in Minnesota," *Wildlife Society Bulletin* 17:121–123.
Fritts, S. H., W. J. Paul, and L. D. Mech. 1985. "Can relocated wolves survive?" *Wildlife Society Bulletin* 13:459–463.
Fritts, S. H., W. J. Paul, L. D. Mech, and D. P. Scott. 1992. *Trends and Management of Wolf-Livestock Conflicts in Minnesota*, U.S. Fish and Wildlife Service, Resource Publication 181, Washington, DC.

Fuller, T. K. 1989. "Population dynamics of wolves in north central Minnesota," *Wildlife Monographs* 105.

Haight, R. G., and L. D. Mech. 1997. "Computer simulation of vasectomy wolf control," *Journal of Wildlife Management* 61:1023-1031.

Harrington, F. H., and L. D. Mech. 1982. "An analysis of howling response parameters useful for wolfpack censusing," *Journal of Wildlife Management* 46:686-693.

Hoogesteijn, R. H., A. H. Hoogesteijn, and E. Mondolfi. 1993. "Jaguar predation and conservation: cattle mortality caused by felines on three ranches in the Venezuelan llanos," *Symposium of the Zoological Society of London* 65:391-407.

Jackson, R. M., G. G. Ahlborn, M. Gurung, and S. Ale. 1996. "Reducing livestock depredation in the Nepalese Himalayas," *Proceedings of the 17th Vertebrate Pest Conference* 17:241-247.

Jorgensen, C. J. 1979. "Bear-sheep interactions, Targhee National Forest," *International Conference on Bear Research and Management* 5:191-200.

Jorgensen, C. J., R. H. Conley, R. J. Hamilton, and O. T. Sanders. 1978. "Management of black bear depredation problems," *Proceedings of the Eastern Workshop on Black Bear Management and Research* 4:297-321.

Kaczensky, P. 1999. "Large carnivore depredation on livestock in Europe," *Ursus* 11:59-72.

Knowlton, F. F., E. M. Gese, and M. M. Jaeger. 1999. "Coyote depredation control: An interface between biology and management," *Journal of Range Management* 52:398-412.

Landa, A., K. Gudvangen, J. E. Swenson, and E. Røskaft. 1999. "Factors associated with wolverine (*Gulo gulo*) predation on domestic sheep," *Journal of Applied Ecology* 36:963-973.

Linnell, J. D. C., R. Aanes, J. E. Swenson, J. Odden, and M. E. Smith. 1997. "Translocation of carnivores as a method for managing problem animals: A review," *Biodiversity and Conservation* 6:1245-1257.

Linnell, J. D. C., J. Odden, M. E. Smith, R. Aanes, and J. E. Swenson. 1999. "Large carnivores that kill livestock: Do 'problem individuals' really exist?" *Wildlife Society Bulletin* 27:698-705.

Minnesota DNR. 2001. *Minnesota Wolf Management Plan.* Minnesota Department of Natural Resources—Division of Wildlife in collaboration with the Minnesota Department of Agriculture.

Mech, L. D. 1970. *The Wolf: The Ecology and Behavior of an Endangered Species.* University of Minnesota Press, Minneapolis.

———. 1974. *Current Techniques in the Study of Elusive Wilderness Carnivores.* Stockholm, Sweden.

———. 1998. "Estimated costs of maintaining a recovered wolf population in agricultural regions of Minnesota," *Wildlife Society Bulletin* 26:817-822.

Mech, L. D., L. G. Adams, T. J. Meier, J. W. Burch, and B. W. Dale. 1998. *The Wolves of Denali.* University of Minnesota Press, Minneapolis.

Mech, L. D., S. H. Fritts, and W. J. Paul. 1988. "Relationship between winter severity and wolf depredations on domestic animals in Minnesota," *Wildlife Society Bulletin* 16:269-272.

Mech, L. D., E. K. Harper, T. J. Meier, and W. J. Paul. 2000. "Assessing factors that may predispose Minnesota farms to wolf depredations on cattle," *Wildlife Society Bulletin* 28:623-629.

Meriggi, A., and S. Lovari. 1996. "A review of wolf predation in southern Europe: Does the wolf prefer wild prey to livestock?" *Journal of Applied Ecology* 33:1561-1571.

Mladenoff, D. J., R. G. Haight, T. A. Sickley, and A. P. Wydeven. 1997. "Causes and implications of species restoration in altered ecosystems," *BioScience* 47:21-31.

Mladenoff, D. J., T. A. Sickley, R. G. Haight, and A. P. Wydeven. 1995. "A regional landscape analysis and prediction of favorable gray wolf habitat in the northern Great Lakes region," *Conservation Biology* 9:279-294.

Mladenoff, D. J., T. A. Sickley, and A. P. Wydeven. 1999. "Predicting gray wolf landscape recolonization: Logistic regression models vs. new field data," *Ecological Applications* 9:37-44.

Mohr, C. O. 1947. "Table of equivalent populations of North American small mammals," *American Midland Naturalist* 37:223-249.

Murdock, W. W. 1969. "Switching in general predators: Experiments on predator specificity and stability of prey populations," *Ecological Monographs* 39:335-354.

Murie, A. 1944. *The Wolves of Mount McKinley*. National Park Service, Washington, DC.

Musiani, M., C. Mamo, L. Boitani, C. Callaghan, C. C. Gates, L. Mattei, E. Visalberghi, S. Breck, and G. Volpi. 2003. "Wolf depredation trends and the use of fladry barriers to protect livestock in western North America," *Conservation Biology*. 17:1538-1547.

Oakleaf, J. K., C. Mack, and D. L. Murray. 2003. "Effects of wolves on livestock calf survival and movements in central Idaho," *Journal of Wildlife Management* 67:299-306.

Ogada, M. O., R. Woodroffe, N. O. Oguge, and L. G. Frank. 2003. "Limiting depredation by African carnivores: The role of livestock husbandry," *Conservation Biology* 17:1521-1530.

Paquet, P. C. 1991. "Winter spatial relationships of wolves and coyotes in Riding Mountain National Park, Manitoba," *Journal of Mammalogy* 72:397-401.

Peters, R. P., and L. D. Mech. 1975. "Scent marking in wolves," *American Scientist* 63:628-637.

Polisar, J. 2000. "Jaguars, pumas, their prey base, and cattle ranching: Ecological perspectives of a management issue." Ph.D. thesis, University of Florida, Gainesville.

Pulliam, H. R. 1988. "Sources, sinks and population regulation," *American Naturalist* 132:652-661.

Quigley, H. B., and A. P. G. J. Crawshaw. 1992. "A conservation plan for the jaguar Panthera onca in the Pantanal region of Brazil," *Biological Conservation* 61:149-157.

Rabinowitz, A. R. 1986. "Jaguar predation on domestic livestock in Belize," *Wildlife Society Bulletin* 14:170-174.

Rajpurohit, K. S. 1998. "Child lifting: Wolves in Hazaribagh, India," *Ambio* 28:163-166.

Rajpurohit, K. S., and P. R. Krausman. 2000. "Human-sloth-bear conflicts in Madhya Pradesh, India," *Wildlife Society Bulletin* 28:393-399.

Robel, R. J., A. D. Dayton, F. R. Henderson, R. L. Meduna, and C. W. Spaeth. 1981. "Relationship between husbandry methods and sheep losses to canine predators," *Journal of Wildlife Management* 45:894-911.

Rothman, R. J., and L. D. Mech. 1979. "Scent marking in lone wolves and newly formed pairs," *Animal Behaviour* 17:750-760.

Sacks, B. N., K. M. Blejwas, and M. M. Jaeger. 1999. "Relative vulnerability of coyotes to removal methods on a northern California ranch," *Journal of Wildlife Management* 63:939-949.

Shivik, J. A., A. Treves, and M. Callahan. 2003. "Nonlethal techniques: Primary and secondary repellents for managing predation," *Conservation Biology*, 17:1531-1537.

Sokal, R. R., and F. J. Rohlf. 1981. *Biometry*. W. H. Freeman, New York.

Stahl, P., and J. M. Vandel. 2001. "Factors influencing lynx depredation on sheep in France: Problem individuals and habitat," *Carnivore Damage Prevention News* 4:6-8.

Stander, P. E. 1990. "A suggested management strategy for stock-raiding lions in Namibia," *South African Journal of Wildlife Research* 20:37-43.

Suminski, H. R. 1982. "Mountain lion predation on domestic livestock in Nevada," *Vertebrate Pest Conference* 10:62-66.

Thiel, R. P. 1993. *The Timber Wolf in Wisconsin: The Death and Life of a Majestic Predator*. University of Wisconsin Press, Madison.

Tiger/Line files™. 1992. Bureau of the Census. Washington, DC.

Tompa, F. S. 1983. "Problem wolf management in British Columbia: Conflict and program evaluation," in *Wolves in Canada and Alaska: Their Status, Biology and Management,* ed. L. N. Carbyn, 112–119. Canadian Wildlife Service, Edmonton, AB.

Treves, A., R. R. Jurewicz, L. Naughton-Treves, R. A. Rose, R. C. Willging, and A. P. Wydeven. 2002. "Wolf depredation on domestic animals: Control and compensation in Wisconsin, 1976–2000," *Wildlife Society Bulletin* 30:231–241.

Treves, A., and K. U. Karanth. 2003. "Human-carnivore conflict and perspectives on carnivore management worldwide," *Conservation Biology.* 17:1491–1499.

Treves, A., and L. Naughton-Treves. 1999. "Risk and opportunity for humans coexisting with large carnivores," *Journal of Human Evolution* 36:275–282.

Treves, A., L. Naughton-Treves, E. L. Harper, D. J. Mladenoff, R. A. Rose, T. A. Sickley, and A. P. Wydeven. 2004. "Predicting human-carnivore conflict: A spatial model derived from 25 years of wolf predation on livestock," *Conservation Biology.* 18:114–125.

Turnbull-Kemp, P. 1967. *The Leopard.* Howard Timmins, Cape Town.

Vogelmann, J. E., S. M. Howard, L. Yang, C. R. Larson, B. K. Wylie, and N. van Driel. 2001. "Completion of the 1990s National Land Cover Data Set for the conterminous United States from Landsat Thematic Mapper data and ancillary data sources," *Photogrammetric Engineering & Remote Sensing* 67:650–652.

Wisconsin DNR. 1999. *Wisconsin Wolf Management Plan,* Wisconsin Department of Natural Resources, Madison, WI.

———. 2002. *Guidelines for Conducting Depredation Control on Wolves in Wisconsin Following Federal Reclassification to "Threatened" Status.* Wisconsin Department of Natural Resources, Madison, WI.

Willging, R., and A. P. Wydeven. 1997. "Cooperative wolf depredation management in Wisconsin," in *Thirteenth Great Plains Wildlife Damage Control Workshop Proceedings,* ed. C. D. Lee and S. E. Hygnstron, 46–51. Kansas State University Agricultural Experiment Station and Cooperative Extension Service, Lincoln, NE.

Wydeven, A. P., R. N. Schultz, and R. P. Thiel. 1995. "Monitoring a recovering gray wolf population in Wisconsin, 1979–1995," in *Ecology and Conservation of Wolves in a Changing World,* ed. L. N. Carbyn, S. H. Fritts, and D. R. Seip, 147–156. Canadian Circumpolar Institute, Occasional Publication No. 35. Edmonton, AB.

Wydeven, A., and J. E. Wiedenhoeft. 2000. "Gray wolf population 1999–2000," *Wisconsin Wildlife Surveys* 10:130–137.

Wydeven, A. P., J. E. Wiedenhoeft, R. N. Schultz, R. P. Thiel, S. R. Boles, and B. E. Kohn. 2002. *Progress Report of Wolf Population Monitoring in Wisconsin for the Period October 2001–March 2002.* Wisconsin Department of Natural Resources, Park Falls, WI.

Young, S. P., and E. A. Goldman. 1944. *The Wolves of North America.* Dover, New York.

CHAPTER 3

Wolves in Rural Agricultural Areas of Western North America: Conflict and Conservation

Marco Musiani, Tyler Muhly, Carolyn Callaghan, C. Cormack Gates, Martin E. Smith, Suzanne Stone, and Elisabetta Tosoni

Gray wolves (*Canis lupus*) prey on all ungulate species present within their distributional range, including domestic ungulates (Young and Goldman 1944; Mech 1970; Meriggi and Lovari 1996). Soon after the introduction of domestic livestock to North America, producers began to experience wolf depredation (Young and Goldman 1944). As you have read in Chapter 1 of this volume, depredation can have a great impact on individual ranchers who suffer losses, and creates severe animosity toward wolves, especially in rural agricultural areas (Mech 1995, 1998). In addition to domestic herbivores, wolves will also kill domestic pets (e.g., hunting dogs, livestock guardian dogs, and other pets; Fritts and Paul 1989; Kojola and Kuittinen 2002; Treves et al. 2002), creating an additional source of conflict between wolves and people. Understanding the factors that influence depredation allows us to consider alternative approaches for managing depredation risk while encouraging wolf conservation.

Historically, wolves were reviled in many areas of North America largely because of wolf-livestock conflicts, and were persecuted by people as a nuisance or pest animal (Young and Goldman 1944; Cluff and Murray 1995). Organized killing of wolves in response to conflicts resulted in their extirpation from most of the lower forty-eight states, with the exception of a wolf population occupying northern Minnesota, at the border with Canada (Mech 1970). In Canada, wolves were extirpated from predominant agricultural areas in the prairie provinces by the early 1900s (Gunson 1992; Hayes and Gunson 1995). Until the 1970s, when the last wolves were killed legally, wolves were repeatedly extirpated from agricultural areas in western North America where recolonization otherwise might have been possible (Fischer 1995). The attitudes and conflicts that existed during the extirpation period are persistent today and can impede wolf recovery.

In this chapter we review conflicts of livestock depredation, compensation, and depredation management for Alberta, Idaho, Montana, and Wyoming, and relevant aspects of wolf recovery, conservation, and legal status. We analyzed depredation data compiled by state and federal agencies over the last twenty years (1982 to 1996 for Canada, 1987 to 2002 for the United States), during which time the wolf was classified as endangered in the U.S. portion of the study area.

Background

In 1973, wolf protection in the lower forty-eight states was embodied in legislation under the Endangered Species Act (ESA) (USFWS 1973). Since then, wolf conservation has received increasing support in North America (Kellert et al. 1996). Proponents of wolf conservation cite evidence of the pivotal role that wolves play in natural ecosystems (Noss et al. 1996; Weaver et al. 1996; Ripple et al. 2001; Hebblewhite et al. 2002; Jedrzejewski et al. 2002). Some sectors of the public advocate complete protection for ethical reasons (Enck and Brown 2002; Williams et al. 2002), while others continue energetic opposition to wolf conservation and recovery (Lohr et al. 1996). Muth and Jamison (2000) explain the motivations for complete protection of wildlife, *sensu* animal rights, which include "urban perceptions of nature," "popularized interpretations of science," "anthropomorphism," and "egalitarianism." Antiwolf sentiments are more strongly held by rural residents in the United States and Canada than by the general public (Bath 1987; Duda et al. 1998; Kellert 1999), reflecting concerns about competition with wolves for large herbivore game species, and real and perceived threats to livestock production (Boyd et al. 1994; Breitenmoser 1998).

The wolf is not considered threatened in Canada because the species, although virtually extirpated from agricultural regions in south-central Canada, has continually occupied much of the remainder of its original range in northern Canada (Hayes and Gunson 1995; Committee on the Status of Endangered Wildlife in Canada 2002). In southern Canada, conflicts with wolves persist, and livestock producers continue to see wolves as a nuisance to livestock production. Ranchers, trappers, and government authorities commonly kill wolves to manage depredation (Bjorge and Gunson 1985; Gunson 1992).

In the mid-1980s, wolves dispersing from Canada began to recolonize northern Montana (Ream et al. 1989; Forbes and Boyd 1997). Management authorities believed that a combination of natural dispersal and reintroduction efforts was needed to establish a viable population in the northwestern United States (Bangs and Fritts 1996). Subsequently, in 1995 and 1996, wildlife officials translocated a number of Canadian wolves into the Greater Yellowstone Area (GYA) and central Idaho (Bangs and Fritts 1996; Bangs et al. 1998). Since then, wolves originating from these populations have expanded into adjacent regions of Idaho, Montana, and Wyoming and now number around 650 individuals (Bangs et al. 1998; USFWS et al. 2003). These recovering populations have come into increasing conflict with livestock operations in the United States (Bangs et al. 1998; USFWS 2003). In addition, there is ongoing conflict in Alberta, where wolf populations have persisted in areas where cattle ranching is a predominant industry (Gunson 1992; Musiani et al. 2003).

In April 2003, the U.S. Fish and Wildlife Service downlisted the wolf from "endangered" to "threatened" under the ESA (USFWS 2003). The reason for downlisting was the improved status and partial recovery of wolf populations in portions of the United States owing to restoration efforts (Bangs et al. 1998). In the northwestern United States, downlisting affected only the northwestern Montana population, as the Yellowstone and Idaho populations were initially classified "experimental, nonessential" (a legal status that allows lethal control of depredating individuals); that classification will be retained for the foreseeable future. However, wolf management may not change significantly as government use of lethal control of wolves has been equal to or greater than in the experimental population. Landowners may now also be issued permits to kill wolves if their livestock is threatened (USFWS 2003). Currently, government officials have the authority to shoot depredating wolves.

Compensation programs are in place to improve tolerance of wolves by livestock producers. The Alberta Conservation Association, a nongovernmental organization (NGO) funded primarily from the province's hunters and anglers, provides funding to the provincial government, which administers the Alberta compensation program (Alberta Conservation Association 2002). In Idaho, Montana, and Wyoming, compensation is provided by a nongovernmental organization, the Defenders of Wildlife (Defenders of Wildlife 2003). As of January 1999, Defenders of Wildlife also began funding proactive depredation management measures. Other antidepredation measures currently employed in the study area include killing or relocation, aversive conditioning, and monitoring the behavior of offending wolves (USFWS et al. 2003).

Depredation Patterns in the Western U.S. and Canada

Effective management of wolf-livestock conflict requires an accurate understanding of the spatial and seasonal patterns of depredation. Our research aims to answer those questions.

Study Area

The case study area consisted of the northwestern U.S. states of Idaho, Montana, and Wyoming, and the Canadian province of Alberta (Figure 3.1). This area includes boreal forest, which is prevalent in northern Alberta. Portions of the study area are occupied by temperate steppe, characterized by agricultural lands and grasslands interspersed with stands of *Populus* spp., with occasional patches of willow (*Salix* spp.). Other portions of the study area encompass the Rocky Mountains, with typical closed to open forests of white and black spruce (*Picea* spp.), subalpine fir (*Abies lasiocarpa*), lodgepole pine (*Pinus contorta*), trembling aspen (*Populus tremuloides*), balsam poplar (*P. balsamifera*), and white birch (*Betula papyrifera*).

Several natural prey species are abundant in parts of the study area, including bison (*Bison bison*), moose (*Alces alces*), elk (*Cervus elaphus*), white-tailed deer (*Odocoileus virginianus*), mule deer (*Odocoileus hemionus*), bighorn sheep (*Ovis canadensis*), and pronghorn antelope (*Antilocapra americana*). Domestic animals, particularly livestock such as cattle, sheep, and horses, are also abundant. Livestock production is an important economic activity on private and public grazing lands.

The region contains both developed areas (towns, agricultural lands, and managed forests) and undeveloped areas (national forests, wilderness areas, and national parks, including Wood Buffalo, Banff/Jasper, Waterton/Glacier, and Yellowstone) (Gunson 1992; Bangs and Fritts 1996).

Methods

We analyzed complaints about wolf depredation on domestic animals in Alberta, Canada, for the period April 1982–April 1996. Alberta Sustain-

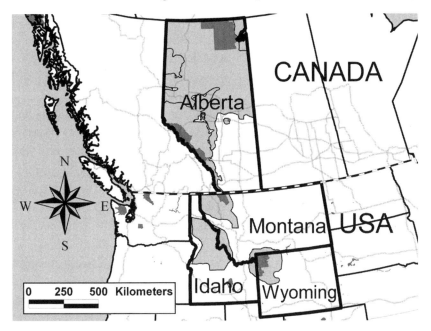

FIGURE 3.1 Study area, including the province of Alberta, Canada, and the northwestern U.S. states of Idaho, Montana, and Wyoming. Light gray denotes ranges for the studied wolf populations. Dark gray denotes national parks.

able Resource Development and Community Development, Government of Alberta, compiled complaint data. Data included instances of: (a) domestic animals that were harassed, killed, or injured by wolves as confirmed by government officials; and (b) wolves killed by government authorities engaged in depredation management. The Alberta database included only information on the occurrences of livestock depredations, with little additional data, but it specified the total number of wolves killed by government authorities in response to these complaints.

We compared Canadian data with those for Wyoming, Idaho, and Montana. The U.S. Fish and Wildlife Service (USFWS) and the USDA Wildlife Services (WS) investigate all depredation complaints. If upon investigation wildlife officials are able to confirm the wolf depredation, then nonlethal, lethal, or a combination of wolf control measures may be implemented under the direction of USFWS (Bangs and Fritts 1996; USFWS et al. 2003).

We compiled data on compensated depredation events originally collected by the USFWS and Defenders of Wildlife from January 1987 to December 2002. In addition to the type of information that was available for Canada, the U.S. database also specified the total number of (a) domestic animals of various age groups that were killed or injured by wolves, and (b) wolves killed or relocated by government authorities.

Using criteria agreed on by the USFWS and WS (Defenders of Wildlife 2003; USFWS et al. 2003), and definitions provided by USDA Wildlife Services, we classified depredation events as "confirmed" (when unequivocal physical evidence was consistent with a wolf attack); "probable" (when available information was consistent with a wolf attack and wolf depredation had been previously confirmed in the area); or "possible" (when insufficient evidence was available to determine the cause of death or injury). Only confirmed events for wolf depredation on domestic animals were included in our analyses. We also obtained information on payments sent to ranchers who had a confirmed or probable loss. Data on the U.S. wolf population were autumn/winter estimates made by the USFWS (USFWS et al. 2003). Depredation data for the United States did not include information on domestic animals harassed by wolves.

Our data represented a minimum estimate for the number of domestic animals killed or injured and for wolves killed. Remains of domestic animals killed by wolves may not be found, or may be found after decomposition or scavenging by other animals. In such cases, assessment of the cause of death is not possible. These detection problems may be particularly relevant in remote areas where livestock are grazed on open ranges (Musiani et al. 2003; Oakleaf et al. 2003).

In Canada or the United States, ranchers are not required to report depredation unless they file a compensation claim for damage. In the United States, all confirmed losses from wolf depredation may be compensated (Wagner et al. 1997). In contrast, losses in Alberta are refunded only for relatively common livestock species, including cattle, sheep, hog, goat, and bison (Alberta Conservation Association 2002). As a consequence, the information available on other livestock (e.g., horse, llama, and alpaca) may be scarce.

Another important difference between Canadian and U.S. data sets is the number of wolf deaths that are reported. In the lower forty-eight states, all wolves killed must be reported in accordance with the Endangered Species Act (USFWS 2003). In Alberta, wolves may be legally killed by hunters (reporting was not required prior to 2000) and registered trappers (reporting required). Landowners may kill wolves without restriction, anywhere on or within 8 km of their property (Gunson 1992).

Analysis

We ran statistical analysis using the Kolmogorov-Smirnov test with reference to Lilliefors's probabilities (Sokal and Rohlf 2000) to determine whether the U.S. data on wolf population size, domestic animals killed and injured, compensation costs, and wolves killed were normally distributed. For normally distributed data, we used analysis of variance (ANOVA) and least-squares simple linear regression to test correlation among yearly numbers. We analyzed separately the period 1987–1994 (natural recolonization period, with wolves present in northwestern Montana only) and the period 1995–2002 (recovery period, with wolves present in Idaho, Montana, and Wyoming).

We evaluated monthly and seasonal trends for both Canadian and American wolf depredation data. In our analysis, we treated each complaint as a single incident, regardless of the numbers of domestic animals affected. To reveal whether depredation events varied by month, we used a Friedman test. We used the Wilcoxon test to evaluate significant increases or decreases from month to month. The latter method allowed us to identify depredation seasons—that is, periods during which depredation occurrences did not change significantly. We used Mann-Whitney U and Kruskal-Wallis tests to compare average seasonal depredation occurrences pairwise and among more than two seasons, respectively (Sokal and Rohlf 2000). All computations were conducted using Statistical Package for the Social Sciences (SPSS Version 10.0). The significance cutoff was $p = 0.05$ for all tests.

Results

Wolf numbers slowly increased in the naturally recolonized area in northwestern Montana between 1987 and 1994 (Figure 3.2a), though the population has undergone temporary declines and long periods of stagnation. The number of domestic animals killed and injured by wolves did not increase during this period ($p = 0.45$), nor did the numbers of domestic animals killed and injured show any relationship with wolf population size during the period ($p = 0.78$). On the other hand, a strong relationship was detected between numbers of domestic animals killed and injured by wolves and numbers of wolves killed by people (Figure 3.2b).

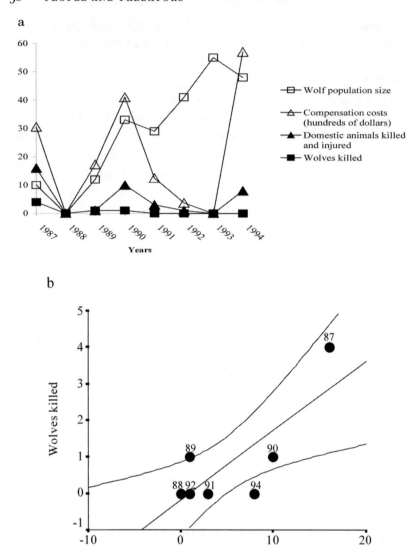

FIGURE 3.2 Trends in wolf population size, domestic animals killed, and injured by wolves, compensation costs, and wolves killed by government authorities in northwestern Montana from 1987 to 1994 (*panel a*). Relationship between the number of domestic animals killed and injured by wolves and wolves killed. Simple linear regression 95% confidence intervals are provided (*panel b*).

Throughout the area in which wolves were reestablished in Idaho, Montana, and Wyoming, wolf numbers increased from 1995 to 2002 in a nearly linear fashion (see Figure 3.3a). In this region, the number of domestic animals killed and injured also did not increase during the period ($p = 0.125$), nor were such depredations linked to yearly wolf population sizes ($p = 0.162$) (see Figure 3.3a). Likewise, numbers of domestic animals killed and injured by wolves were strongly related to numbers of wolves killed by people (see Figure 3.3b).

In Canada and the United States, a similar proportion (13% for each country) of submitted compensation claims were for domestic animals injured by wolves. In Alberta, from 1982 to 1996, wolves killed cattle on 756 occasions, dogs on 77 occasions, and sheep on 61 occasions. Wolves also reportedly killed horses and other domestic animal species on 40 and 87 occasions, respectively. In Idaho, Montana, and Wyoming, from 1987 to 2002, wolves killed cattle on 164 occasions (125 attacks on calves, 19 on yearlings, and 20 on adults); sheep on 88 occasions (55 attacks on adults and 33 on lambs); and dogs on 27 occasions.

Sheep were killed more often in the United States than in Alberta (31% versus 6% of all occurrences). Sheep represented 69% of all animals killed by wolves in the United States (data on numbers of animals depredated were not available for Canada). Similarly, 68% of animals injured in the United States were sheep. Calves clearly constituted the majority of U.S. cattle killed or injured by wolves (80%; Table 3.1). The Canadian database included figures for domestic animals harassed by wolves and showed that cattle were harassed on 288 occasions, dogs on 85 occasions, and all other species on 74 occasions (see Table 3.1).

We have no data on compensation trends for the Canadian portion of the study area, but throughout Idaho, Montana, and Wyoming (1987 to 2002), annual compensation payments were significantly related to the numbers of domestic animals killed and injured by wolves, with payments increasing from 1987 to 2002 (see Figures 3.2 and 3.3). Our data demonstrated that the number of wolves killed in management control actions was also consistently related to the number of domestic animals killed or injured by wolves (see Figures 3.2 and 3.3). Thus, our findings indicate that the damage compensation and lethal wolf control are primarily used in response to wolf depredation.

Wolf depredation on livestock appears to follow a seasonal pattern. From 1987 to 2002, in Idaho, Montana, and Wyoming, occurrence of

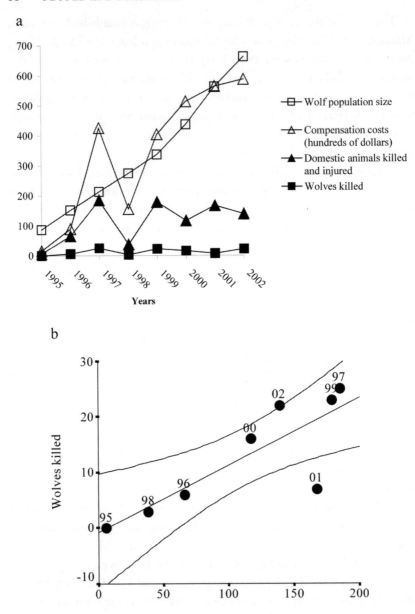

FIGURE 3.3 Trends in wolf population size, domestic animals killed and injured by wolves, compensation costs, and wolves killed by government authorities in Idaho, Montana, and Wyoming from 1995 to 2002 (after wolf reintroductions) (*panel a*). Relationship between the number of domestic animals killed and injured by wolves and wolves killed. Simple linear regression 95% confidence intervals are provided (*panel b*).

TABLE 3.1
The number of occurrences and number of individuals (in parentheses) harassed, injured, or killed by wolves in Alberta, Canada (*top section*), or in Idaho, Montana, and Wyoming, USA (*bottom section*). Wolves refer to the number of wolf control actions and the number of individual wolves killed in the two areas.

Alberta 1982–1996	Harassment Events	Injury Events	Kill Events
Cattle	288	170	756
Dogs	85	25	77
Sheep	5	4	61
Horses	31	13	40
Chickens	5		19
Goats			13
Bison	5		9
Geese	1		4
Turkeys			3
Other	27	4	39
Total	447	216	1021
Wolves			246 (795)

Idaho, Montana, and Wyoming: 1987–2002	Harassment Events	Injury Events (no. injured)	Kill Events (no. killed)
Cattle			
Calves		10 (4)	125 (178)
Yearlings		4 (11)	19 (24)
Adults		4 (5)	20 (25)
Sheep			
Lambs			33 (302)
Adults		9 (63)	55 (300)
Dogs		8 (9)	27 (38)
Horses			1 (3)
Total		35 (92)	280 (870)
Wolves			62 (108)

depredation events varied significantly by month. Our data revealed significant increases in number of depredations from February to March and significant decreases from October to November (Figure 3.4). Depredation occurrences did not change significantly in two uniform seasons lasting from March to October and from November to February, respectively. More depredations occurred between March and October than between November and February. Thus, we could identify a two-season pattern in the occurrence of depredations in the United States, which is comprised of

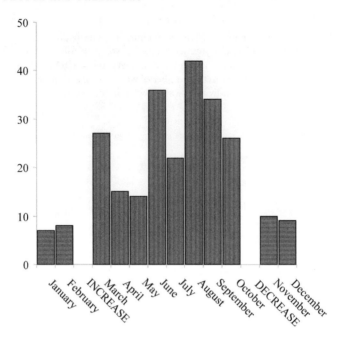

FIGURE 3.4 Monthly occurrences of wolf depredation on domestic animals in Idaho, Montana, and Wyoming from 1987 to 2002. Hatching indicates low or medium-high depredation seasons (horizontal and vertical lines, respectively).

a "low depredation season" (November to February) and a "medium-high depredation season" (March to October) (see Figure 3.4).

In Alberta, depredation occurrences also varied by month (Figure 3.5). Our data revealed decreases in depredations from January to February; increases from April to May; and decreases again from September to October (see Figure 3.5). Fewer depredations occurred between February and April than between May and September. The period from October to January had fewer depredations than May to September, but more than February to April. Thus, we identified a three-season pattern in the occurrence of depredations in Canada, which is comprised of a distinctive "medium depredation season" (October to January), a "low depredation season" (February to April), and a "high depredation season" (May to September).

Discussion

The expansion of the wolf population since reintroductions in Yellowstone National Park and central Idaho without a related increase in the number

3. Wolves in Rural Agricultural Areas of Western North America   63

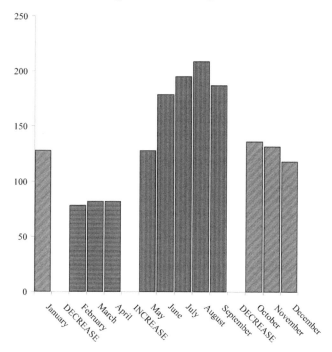

FIGURE 3.5   Monthly occurrences of wolf depredation on domestic animals in Alberta, Canada from 1982 to 1996. Hatching indicates low, medium, or high depredation seasons (horizontal, diagonal and vertical lines respectively).

of domestic animals killed or injured is contrary to initial predictions (Bangs et al. 1998; Musiani et al. 2003). In fact, in recent years there was no significant relationship between numbers of wolves present and the number of domestic animals killed or injured by wolves. A potential explanation for the slower increase in the number of domestic animals depredated is that lethal control of problem wolves (*sensu* Linnell et al. 1999) by government authorities eliminated individuals or packs from the population that had "learned" to take livestock. Potential repeat offenders may be selected against, and these offenders may not have the opportunity to teach their offspring or other pack members how to kill livestock, thus reducing further depredations. This would require livestock depredation to be a learned behavior that can be retained by individuals and propagated among wolves in a pack or population.

Linnell et al. 1999 argued against the idea of problem individuals existing in predator populations. Further research is urgently needed to test whether learning and propagation are possible in wolves. A second, nonexclusive possibility is that there is a limited number of areas where environmental conditions predispose livestock to depredation. This proposition requires study of the biophysical factors that contribute to livestock depredation at the landscape level (see approach by Mech 1998).

The higher proportion of attacks on sheep in the United States compared with Alberta is likely related to the differential availability of sheep. In Alberta, there are few sheep operations in proximity to wolf range (Musiani et al. 2003), whereas in the United States many sheep operations are within wolf range, particularly in Idaho and Wyoming (Bangs et al. 1998). The high prevalence of calves recorded is in agreement with previous studies showing that wolves disproportionately select younger cattle (Bjorge and Gunson 1985; Fritts et al. 1992; Oakleaf et al. 2003). There may also be spatial and temporal variability in the availability and vulnerability of young cattle that we could not infer from the available data.

The effects of harassment on livestock are poorly understood. Harassment may cause reduced weight gain, such as observed for white-tailed deer (Berger et al. 2001), though an extensive study in Salmon, Idaho, on cattle raised with a high degree of persistent wolf predation did not find any significant weight loss (Oakleaf et al. 2003). Our results on cattle harassed by wolves in Canada reflect the prevalence of cattle among domestic animals present. The harassment of dogs may reflect that wolves and dogs are closely related species (Leonard et al. 2002) and thus interact aggressively toward each other (Coppinger and Coppinger 1995).

In the U.S. portion of the study area, it is possible that increasing collaboration between government, NGOs, and ranchers is improving animal husbandry practices and reducing the number of domestic animals depredated. Some ranchers are actively participating in wolf depredation management by monitoring wolf movements close to farms and relocating livestock herds when wolves are present (Defenders of Wildlife 2003). Similar cooperation may also be developing in Canada, where a few experiments with nonlethal methods have occurred (see Musiani et al. 2003). Alternatively, the combination of lethal and nonlethal management approaches currently in place in the United States (e.g., killing and relocation of offending wolves, compensation, aversive conditioning, fencing, rancher education, etc.) may be limiting the impact of depredation through complex interactions involving multiple factors. In short, we cannot specifically identify which individual approach, or combination of techniques, is most important in mitigating depredation risk.

Seasonal Depredation Patterns

In Idaho, Montana, and Wyoming the grazing period typically lasts from May to October (Oakleaf et al. 2003). Although this could explain increased depredation during May to October, it fails to explain the spike in March, the beginning of the medium-high depredation season. One possible explanation is that calving often occurs during midspring (Cole 1966). The magnitude of wolf depredation on calves (see above) together with the timing of the grazing season helped explain the two-season pattern of wolf depredation in the United States (see Figure 3.4).

In Alberta, the grazing period varies among years and areas as a result of variation in weather conditions. However, as a general pattern, the typical grazing period starts in May. In most areas, livestock is not grazed after mid-October (Barry Adams, Range Management Specialist, Sustainable Resource Development, Government of Alberta, pers. comm.). In a few areas, grazing may be continued to December/January. February through April is the period when the fewest livestock are grazed (Lodge 1970). Thus, also in Canada, grazing practices could help explain the seasonal pattern of wolf depredation (see Figure 3.5).

In both countries, the peak in depredation occurred in August, which is the month during which the protein demands for a wolf pack might also be at a peak, owing to demand for nurturing pups (Mech 1970). Depredation similarly decreased during the winter in both the United States and Canada. The widespread practice of supplemental feeding in winter may reduce depredation risk because herds are often relocated to pastures closer to main farm buildings and there is an increased presence of people near the pastures. Finally, livestock are concentrated in dense herds for supplemental feeding during the winter, thus reducing the vulnerability of individuals to attack by predators (i.e., evolutionary advantage of herds; Lingle 2001).

Vigilance of ranchers at different times of the year may also influence the pattern of depredation. Alberta ranchers have had to deal with wolf depredation almost continuously (Gunson 1992), whereas U.S. ranchers have had a gap of a few generations in dealing with depredation because wolves were extirpated (Mech 1970; Fischer 1995; Bangs and Fritts 1996). The low depredation season in Alberta during late winter/early spring coincides with the beginning of the medium-high depredation season in the United States. This difference may be due to better vigilance by Alberta ranchers during calving, which typically occurs in pastures close to ranch buildings and is closely monitored by the ranchers.

Possible Solutions

Regardless of the legal status of wolves, there is a similar need in Canada and the United States to develop a range of alternatives for simultaneously managing wolf depredation and fostering wolf conservation. Several nonlethal and lethal methods have been employed to mitigate conflicts in different regions of the world.

Financial compensation is one of the most common methods for mitigating wolf depredation conflicts. Payments are typically administered and funded by government authorities and/or private organizations (Mech 1995, 1998; Treves et al. 2002). Compensation payments can help to relieve the economic burden of wolf depredation and increase tolerance of wolves by rural communities (Wagner et al. 1997; Ciucci and Boitani 1998; Phillips and Smith 1998). Compensation programs funded by private organizations are less likely to be criticized by taxpayers (Mech 1995, 1998). However, private funding may not be guaranteed in the long term. Compensation may also provide a perverse subsidy in the sense that livestock managers may reduce efforts to protect their stock. Some authors suggest that compensation should be coupled with strong policies favoring management actions that reduce depredation risk (Cozza et al. 1996; Poulle et al. 1997; Ciucci and Boitani 1998).

Surveillance of livestock herds is the most traditional nonlethal method used by ranchers. It is a simple, but labor-intensive technique (Mysterud et al. 1996a, 1996b) that can be expensive if the producer employs staff to watch over livestock (Bjorge and Gunson 1985; Ciucci and Boitani 1998).

The use of guardian dogs to protect livestock, particularly sheep, from predators is widespread in southern Europe and Asia (Hansen and Smith 1999; Smith et al. 2000a). In North America, the use of guardian dogs became more common when poison control of coyotes was banned in the 1970s (Coppinger and Coppinger 1995, 2001). In spite of additional costs in raising and maintaining dogs, their use is now fairly common in the United States, and although they are most effective against smaller predators, they have also been used successfully in disrupting depredation by bears and wolves (Smith et al. 2000a).

Fencing techniques are also used to deter predators such as wolves and bears. Electric fences, or combinations of wire mesh and electric fences, have proven to be particularly effective (Gipson and Paul 1994). However, permanent predator-proof fencing is of limited use when livestock are kept in large enclosures, because such fences are costly to build and maintain (Smith and Christiansen, in prep.). In addition, permanent fences are generally not portable and therefore of little use when livestock are moved from one pasture to the other. In western North America, livestock are often kept in big pastures or are grazed on open ranges during the spring, summer, and fall months (Lodge 1970; Oakleaf et al. 2003), making fencing impractical in many instances. However, the use of portable and permanent night pens has been successful in stopping chronic depredations in the Nine Mile Valley in Montana and the Sawtooth National Recreation Area in Idaho, and warrants consideration in favorable situations.

Capturing and translocating individual wolves involved in depredations is another nonlethal option. However, translocation programs are costly and labor-intensive, and may require specialized veterinary skills for sedating and transporting wolves (Fritts 1982; Fritts et al. 1985; Linnell et al. 1997). Translocation may cause indirect mortality resulting from dominance fights with resident wolves residing in release areas.

Another nonlethal method, sometimes combined with translocation, involves "training" wolves not to prey on livestock through aversive conditioning (Fritts et al. 1992; Smith et al. 2000b; Bangs and Shivik 2001). One technique is to fit captured wolves with electric shock collars and shock them as they approach livestock. These wolves are then translocated back into the wild, and those that return to attacking livestock are subsequently lethally removed.

Another aversive conditioning technique is to feed wolves livestock carcasses laced with chemicals that will make the wolves sick and hopefully keep them from eating that type of meat again (Smith et al. 2000b). The major drawbacks of both translocation and aversive conditioning are high costs and inconsistent results, often with only short-term effects, if any (Scott et al. 1989; Linnell et al. 1996; Smith et al. 2000b).

Lethal methods involve killing depredating individuals or entire packs of wolves (Cluff and Murray 1995; Mech 2001). Lethal control is controversial and has been criticized as ineffective and inhumane (Struzik 1993;

Haber 1996; Berg 1998). Evidence suggests that lethal control can be effective at reducing depredation (Fritts 1982; Tompa 1983; Bjorge and Gunson 1985). However, effectiveness of lethal control depends on various parameters that are often difficult to manage, including wolf population density, size and age structure of packs, and age and social status of wolves killed. Implementing lethal control methods, including helicopter gunning and exhaustive trapping efforts, can be very expensive and elevate risk to human safety.

Given these uncertainties, there is an urgent need for thorough reevaluation of lethal control programs and their effects on both depredation by wolves and wolf population survival. Lethal control has historically been the leading cause of extirpation of wolves in the absence of a conservation objective (Mech 1970). Consideration of alternatives to the controversial practice of killing depredating wolves will require an improved understanding of factors that influence depredation risk and which are also amenable to management (Mech 1995, Musiani et al. 2003).

Conclusion

In Canada, the lack of legislation to protect wolves (because of the relative stability of wolf populations; Committee on the Status of Endangered Wildlife in Canada 2002) provides liberal opportunities for killing wolves that cause livestock depredation. Control actions in Canada could negatively impact recovery efforts in the United States if emigrants or immigrants are killed in Canada (USFWS 2003). In our study area, there is an important link between the two countries because wolves move between Alberta, Montana, and northern Idaho (Forbes and Boyd 1997). To support recovery efforts in the United States, there should be more cooperation and sharing of knowledge between the two countries to reduce conflicts arising from livestock depredation.

Our results are encouraging because the number of domestic animals killed and injured by wolves is not increasing in proportion to the increase in the wolf population, as was initially predicted for the United States (Bangs et al. 1998). We evaluated a variety of explanations for these trends that might shed light on how the depredation problem could be mitigated. Our analysis of the seasonal patterns of wolf depredation suggests the need to protect domestic animals at specific vulnerable times of the year (see Figures 3.4 and 3.5). However, we must be cautious in predicting future trends. Thus far, predation in the United States has primarily occurred within the original recovery program areas (Bangs et al. 1998; USFWS et al. 2003). Geographic expansion of the wolf population could lead to increased depredation outside of the recovery area, where environmental and socioeconomic contexts may be different.

Future research should focus on determining areas of high depredation occurrence, and identifying which factors make such areas differentially susceptible to depredation. This knowledge could then be used to assess depredation risk at the landscape level and to plan suitable management actions in areas at higher risk for livestock depredation.

Efforts should continue to evaluate objectively the effectiveness of various management approaches in reducing depredation. Lethal control may be effective in some circumstances, but it is contentious because large sectors of the public oppose it, and it may not prevent future depredations if wolves recolonize the same area and are exposed to the same conditions that led to the initial depredation. In addition to techniques traditionally used to protect livestock or to kill wolves, described above, new methods should be tested that might reduce the risk of depredation while providing for effective wolf conservation.

For example, our group is testing a management technique called fladry (barriers made by surrounding an area with ropes that have 18 × 2 inch flags attached every couple of feet), which is designed to repel wolves (Musiani and Visalberghi 2001; Musiani et al. 2003; Figure 3.6). During experiments conducted in captivity, fladry prevented wolves from accessing their daily food ration. Fladry also prevented access by wild wolves to baited sites, where we attempted to lure wolves to road kills. In Alberta and Idaho, we also set fladry around cattle pastures (see Figure 3.6). Wolves approached on various occasions, but did not cross fladry, nor were any cattle killed.

In Idaho, wolves avoided the fladry for 63 days but then crossed the fladry and killed livestock. It was unclear if the fladry was tangled and therefore ineffective at the time the wolves crossed it. However, helicopter hazing could not force the wolves back through the fladry even after they penetrated the line and killed livestock, which suggests that the wolves still strongly avoided the fladry. Meanwhile wolves did kill cattle on neighboring ranches during the trials, as well as before and after the trials on the tested ranches.

Our results suggested that fladry has potential for protecting livestock

FIGURE 3.6 Fladry barrier (flags hanging from a rope stretched above the ground) employed to protect livestock from wolves on a Canadian ranch. Photo by M. Musiani.

from wolves, but that wild wolves can switch to adjacent available livestock herds. Thus, fladry may offer a cost-effective tool for reducing livestock depredation on a local, short-term basis. However, fladry may be ineffective when no alternative food sources are available to wolves—that is, we do not advise using fladry on gigantic areas and/or for long periods of time if enough wild prey is not available.

Historically, people quickly eradicated wolves from large portions of the study area (Gunson 1992; Fischer 1995). It is possible that continued conflicts with people, particularly livestock depredation, could again result in wolf extirpation, at least from certain regions and for short time periods. Although the social context and the values assigned to wolves are changing, particularly among urban dwellers (Kellert et al. 1996), the study area is still characterized by extensive livestock production and a social climate that commonly opposes wolf recovery. Wolf survival in rural agricultural areas is disproportionately dependent on the actions of human residents whose livelihoods depend on the productivity of the landscape they share. Conservationists are challenged to work with ranchers and others experiencing depredation to improve methods for mitigating impacts and increasing tolerance of wolves.

Acknowledgments

We acknowledge the ranching communities of Alberta, Idaho, Montana, and Wyoming; Alberta Beef Producers, Alberta Sustainable Resource Development and Community Development, The Bailey Wildlife Foundation Wolf Compensation Trust, Defenders of Wildlife, USDA Wildlife Services, USFWS, Ed Bangs, Joe Bews, Nina Fascione, Laura Jones, Timothy Kaminsky, Terry Mack, Charles Mamo, Carter Niemeyer, Carol Powell, and Gary Sargent. We acknowledge support from the Alberta Conservation Association, Alberta Ecotrust, Calgary Foundation, Calgary Zoo, Humane Society U.S., TD Friends of the Environment foundation, and Yellowstone to Yukon Conservation Initiative. MM was supported by Honourary Killam, NSERC, Canada, and Consiglio Nazionale delle Ricerche, Italy. ET was supported by the University of Rome, Italy.

Literature Cited

Alberta Conservation Association. 2002. *2001/02 Annual Report*. ACA, Edmonton, AB. www.ab-conservation.com/about_us/reports_publications/Annual_Report_2001_2002.pdf.

Bangs, E. E., and S. H. Fritts. 1996. "Reintroducing the gray wolf to central Idaho and Yellowstone National Park," *Wildlife Society Bulletin* 24:402–413.

Bangs E. E., S. H. Fritts, J. A. Fontaine, D. W. Smith, K. M. Murphy, C. M. Mack, and C. C. Niemeyer. 1998. "Status of gray wolf restoration in Montana, Idaho, and Wyoming," *Wildlife Society Bulletin* 26:785–798.

Bangs, E. E., and J. Shivik. 2001. "Managing wolf conflict with livestock in the northwestern United States," *Carnivore Damage Prevention News* 3:2–5.

Bath, A. J. 1987. "Attitudes of various interest groups in Wyoming toward wolf reintroduction in Yellowstone National Park." M.A. thesis, University of Wyoming.

Berg, K. A. 1998. "The future of the wolf in Minnesota: Control, sport or protection?" *Proceedings of the Defenders of Wildlife's Restoring the Wolf Conference*, Seattle, WA.

Berger, J., J. E. Swenson, and I. L. Persson. 2001. "Recolonizing carnivores and naïve prey: Conservation lessons from Pleistocene extinctions," *Science* 291:1036–1039.

Bjorge, R. R., and J. R. Gunson. 1985. "Evaluation of wolf control to reduce cattle predation in Alberta," *Journal of Range Management* 38:483–487.

Boyd, D. K., R. R. Ream, D. H. Pletscher, and M. W. Fairchild. 1994. "Prey taken by colonizing wolves and hunters in the Glacier National Park area," *Journal of Wildlife Management* 58:289–295.

Breitenmoser, U. 1998. "Large predators in the Alps: The fall and rise of man's competitors," *Biological Conservation* 83:279–289.

Ciucci, P., and L. Boitani. 1998. "Wolf and dog depredation on livestock in central Italy," *Wildlife Society Bulletin* 26:504–514.

Cluff, H. D., and D. L. Murray. 1995. "Review of wolf control methods in North America," in *Ecology and Conservation of Wolves in a Changing World*, ed. L. N. Carbyn, S. H. Fritts, and

D. R. Seip, 491–607. Canadian Circumpolar Institute, Occasional Publication no. 35. University of Alberta, Edmonton.

Cole, H. H. 1966. *Introduction to Livestock Production.* W. H. Freeman, San Francisco.

Committee on the Status of Endangered Wildlife in Canada. 2002. COSEWIC Species Database. Date published: January 21, 2002, Modified: September 25, 2002. www.cosewic.gc.ca/eng/sct1/searchform_e.cfm.

Coppinger, R., and L. Coppinger. 1995. "Interactions between livestock guarding dogs and wolves," in *Ecology and Conservation of Wolves in a Changing World,* ed. L. N. Carbyn, S. H. Fritts, and D. R. Seip, 523–526. Canadian Circumpolar Institute, Occasional Publication no. 35. University of Alberta, Edmonton.

———. 2001. *Dogs: A New Understanding of Canine Origin, Behavior and Evolution.* University of Chicago Press, Chicago.

Cozza, K., R. Fico, M. L. Battistini, and E. Rogers. 1996. "The damage-conservation interface illustrated by predation on domestic livestock in central Italy," *Biological Conservation* 78:329–336.

Defenders of Wildlife. 2003. www.defenders.org/wildlife/new/wolves.html.

Duda, M. D., S. J. Bissell, and K. C. Young. 1998. *Wildlife and the American Mind: Public Opinion and Attitudes Toward Fish and Wildlife Management.* Responsive Management, Harrisonburg, VA.

Enck, J. W., and T. L. Brown. 2002. "New Yorkers' attitudes toward restoring wolves to the Adirondack Park," *Wildlife Society Bulletin* 30:16–28.

Fischer, H. 1995. *Wolf Wars: The Remarkable Inside Story of the Restoration of Wolves to Yellowstone.* Falcon Press, Helena, MT.

Forbes, S. H., and D. K. Boyd. 1997. "Genetic structure and migration in native and reintroduced Rocky Mountain wolf populations," *Conservation Biology* 11:1226–1234.

Fritts, S. H. 1982. *Wolf Depredation on Livestock in Minnesota.* U.S. Fish and Wildlife Service, Washington, DC, Resource Publication 145.

Fritts, S. H., and W. J. Paul. 1989. "Interactions of wolves and dogs in Minnesota," *Wildlife Society Bulletin* 17:121–123.

Fritts, S. H., W. J. Paul, and L. D. Mech. 1985. "Can relocated wolves survive?" *Wildlife Society Bulletin* 13:459–463.

Fritts, S. H., W. J. Paul, L. D. Mech, and D. P. Scott. 1992. *Trends and Management of Wolf-Livestock Conflicts in Minnesota.* U.S. Fish and Wildlife Service, Washington, DC, Resource Publication 181.

Gipson, P. S., and W. J. Paul. 1994. *Wolves—Prevention and Control of Wildlife Damage.* United States Department of Agriculture, Great Plains Agricultural Council. University of Nebraska Press, Lincoln.

Gunson, J. R. 1992. "Historical and present management of wolves in Alberta," *Wildlife Society Bulletin* 20:330–339.

Haber, G. C. 1996. "Biological, conservation, and ethical implications of exploiting and controlling wolves," *Conservation Biology* 10:1068–1081.

Hansen, I., and M. E. Smith. 1999. "Livestock-guarding dogs in Norway—Part II: Different working regimes," *Journal of Range Management* 52:312–316.

Hayes, R. D., and J. R. Gunson. 1995. "Status and management of wolves in Canada," in *Ecology and Conservation of Wolves in a Changing World,* ed. L. N. Carbyn, S. H. Fritts, and D. R. Seip, 21–33. Canadian Circumpolar Institute, Occasional Publication no. 35, Edmonton, AB.

Hebblewhite, M., D. H. Pletscher, and P. C. Paquet. 2002. "Elk population dynamics in areas with and without predation by recolonizing wolves in Banff National Park, Alberta," *Canadian Journal of Zoology* 80:789–799.

Jedrzejewski, W., K. Schmidt, J. Theuerkauf, B. Jedrzejewska, N. Selva, K. Zub, and L. Szymura. 2002. "Kill rates and predation by wolves on ungulate populations in Bialowieza Primeval Forest (Poland)," *Ecology* 83:1341–1356.

Kellert, S. R., 1999. "The public and the wolf in Minnesota," Report for the International Wolf Center. New Haven, CT.

Kellert, S. R, M. Black, C. R. Rush, and A. J. Bath. 1996. "Human culture and large carnivore conservation in North America," *Conservation Biology* 10:977–990.

Kojola, I., and J. Kuittinen. 2002. "Wolf attacks on dogs in Finland," *Wildlife Society Bulletin* 30:498–501.

Leonard, J. A, R. K. Wayne, J. Wheeler, R. Valadez, S. Guillen, and C. Vilà. 2002. "Ancient DNA evidence for Old World origin of New World dogs," *Science* 298:1613–1616.

Lingle, S. U. S. 2001. "Anti-predator strategies and grouping patterns in white-tailed deer and mule deer," *Ethology* 107:295–314.

Linnell, J. D. C., R. Aanes, J. E. Swenson, J. Odden, and M. E. Smith. 1997. "Translocation of carnivores as a method for managing problem animals: A review," *Biodiversity and Conservation* 6:1245–1257.

———. 1999. "Large carnivores that kill livestock: Do 'problem individuals' really exist?" *Wildlife Society Bulletin* 27:698–705.

Linnell, J. D. C., M. E. Smith, J. Odden, P. Kaczensky, and J. E. Swenson. 1996. *Carnivores and Sheep Farming in Norway. 4. Strategies for the Reduction of Carnivore-Livestock Conflicts: A Review*. Norwegian Institute for Nature Research, Project Report (Oppdragsmelding) No. 443.

Lodge, R. W. 1970. "Complementary grazing system for the northern great plains [MAS1]," *Journal of Range Management* 23:268–271.

Lohr, C., W. B. Ballard, and A. J. Bath. 1996. "Attitudes toward gray wolf reintroductions to New Brunswick," *Wildlife Society Bulletin* 24:414–420.

Mech, L. D. 1970. *The Wolf: The Ecology and Behavior of an Endangered Species*. University of Minnesota Press, Minneapolis.

———. 1995. "The challenge and opportunity of recovering wolf populations," *Conservation Biology* 9:270–278.

———. 1998. "Estimated costs of maintaining a recovered wolf population in agricultural regions of Minnesota," *Wildlife Society Bulletin* 26:817–822.

———. 2001. "Managing Minnesota's recovered wolves," *Wildlife Society Bulletin* 29:70–77.

Meriggi, A., and S. Lovari. 1996. "A review of wolf predation in southern Europe: Does the wolf prefer wild prey to livestock?" *Journal of Applied Ecology* 33:1561–1571.

Musiani, M., C. Mamo, L. Boitani, C. Callaghan, C. Gates, L. Mattei, E. Visalberghi, S. Breck, and G. Volpi. 2003. "Wolf depredation trends and the use of fladry barriers to protect livestock in western North America," *Conservation Biology* 17:1538–1547.

Musiani M., and E. Visalberghi. 2001. "Effectiveness of fladry on wolves in captivity," *Wildlife Society Bulletin* 29:91–98.

Muth, R. M., and W. V. Jamison. 2000. "On the destiny of deer camps and duck blinds: The rise of the animal rights movement and the future of wildlife conservation," *Wildlife Society Bulletin* 28:841–851.

Mysterud, I., A. O. Gautestad, and I. Mysterud. 1996a. *Rovvilt og sauenæring i Norge. 6.*

Kommentarer til gjeting som forebyggende tiltak (Carnivores and Sheep Farming in Norway. 6. Comments on Shepherding as a Preventive Measure). Project Report, Norwegian Institute for Nature Research, Trondheim, Norway. (Norwegian with English abstract.)

Mysterud, I., J. E. Swenson, J. C. D. Linnell, A. O. Gautestad, I. Mysterud, J. Odden, M. E. Smith, R. Aanes, and P. Kaczensky. 1996b. *Rovvilt og sauenæring i Norge: Kunnskapsoversikt og evaluering av forebyggende tiltak. Sluttrapport (Carnivores and Sheep Farming in Norway: A Survey of the Information and Evaluation of Mitigating Measures).* Final report, Norwegian Institute for Nature Research, Trondheim, Norway. (Norwegian with English abstract.)

Noss, R. F., H. B. Quigley, M. G. Hornocker, T. Merrill, and P. C. Paquet. 1996. "Conservation biology and carnivore conservation in the Rocky Mountains," *Conservation Biology* 10:949–963.

Oakleaf, J. K., C. Mack, and D. L. Murray. 2003. "Effects of wolves on livestock calf survival and movements in central Idaho," *Journal of Wildlife Management* 67: 299–306.

Phillips, M. K., and D. W. Smith. 1998. "Gray wolves and private landowners in the Greater Yellowstone Area," *Transactions of the North American Wildlife and Natural Resources Conference* 63:443–450.

Poulle, M. L., L. Carles, and B. Lequette. 1997. "Significance of ungulates in the diet of recently settled wolves in the Mercantour mountains (southeastern France)," *Revue d'Ecologie—La Terre et la Vie* 52:357–368.

Ream, R. R., M. W. Fairchild, D. K. Boyd, and A. J. Blakesley. 1989. "First wolf den in western U.S. in recent history," *Northwestern Naturalist* 70:39–40.

Ripple, W. J., E. J. Larsen, R. A. Renkin, and D. W. Smith. 2001. "Trophic cascades among wolves, elk and aspen on Yellowstone National Park's northern range," *Biological Conservation* 102:227–234.

Scott, J. M., J. W. Carpenter, and C. Reed. 1989. "Translocation as a species conservation tool: Status and strategy," *Science* 245:477–480.

Smith, M. E., and F. O. Christiansen. In prep. "The cost and efficacy of using herding and livestock guardian dogs to reduce sheep depredation in Lierne, Norway."

Smith, M. E., J. D. C. Linnell, J. Odden, and J. E. Swenson. 2000a. "Review of methods to reduce livestock depredation: I. Guardian animals," *Acta Agriculturae Scandinavica,* Sect. A, Animal Sciences, 50:279–290.

———. 2000b. "Review of methods to reduce livestock depredation: II. Aversive conditioning," *Acta Agriculturae Scandinavica* Sect. A, Animal Sciences, 50:304–315.

Sokal, R. S., and J. R. Rohlf. 2000. *Biometry.* W. H. Freeman, New York.

Struzik, E. 1993. "Managing the competition: Wolves are at the center of a wildlife management controversy," *Nature Canada* 22:22–27.

Tompa, F. S. 1983. *Status and Management of Wolves in British Columbia.* Canadian Wildlife Service Report Series 45:112–119.

Treves A., R. R. Jurewicz, L. Naughton-Treves, R. A. Rose, R. C. Willging, and A. P. Wydeven. 2002. "Wolf depredation on domestic animals in Wisconsin, 1976–2000," *Wildlife Society Bulletin* 30:231–241.

USFWS. 1973. *Endangered Species Act, ESA.* Enacted on December 28, 1973. Government of the U.S., Washington, DC. http://endangered.fws.gov/esa.html.

———. 2003. "Endangered and threatened wildlife and plants; final rule to reclassify and remove the gray wolf from the list of endangered and threatened wildlife in portions of the conterminous United States; establishment of two special regulations for threatened gray wolves; final and proposed rules," *Federal Register,* 68, no. 62, 50 CFR, Part 17.

USFWS, Nez Perce Tribe, National Park Service, and USDA Wildlife Services. 2003. *Rocky Mountain Wolf Recovery, 2002 Annual Report,* T. Meier, ed. USFWS, Ecological Services, 100 N Park, Suite 320, Helena MT. http://mountain-prairie.fws.gov/wolf/annualrpt02.

Wagner, K. K., R. H. Schmidt, and M. R. Conover. 1997. "Compensation programs for wildlife damage in North America," *Wildlife Society Bulletin* 25:312–319.

Weaver, J. L., P. C. Paquet, and L. F. Ruggiero. 1996. "Resilience and conservation of large carnivores in the Rocky Mountains," *Conservation Biology* 10:964–976.

Williams, C. K., G. Ericsson, and T. A. Heberlein. 2002. "A quantitative summary of attitudes toward wolves and their reintroduction (1972–2000)," *Wildlife Society Bulletin* 30:575–584.

Young, S. P., and E. A. Goldman. 1944. *The Wolves of North America.* Dover, New York.

PART 2

Coexistence in Developed Landscapes

As we have seen in Part 1, coexistence between humans and carnivores in rural landscapes frequently depends on the extent to which conflicts between carnivores and livestock, a critically important rural resource, can be avoided or mediated. The chapters in Part 1 demonstrate that research into the patterns and causes of predator-livestock interactions, alteration of husbandry practices, and implementation of proactive measures to keep predators away from livestock can help to reduce these conflicts.

As landscapes become more developed, however, new sets of challenges arise. Urbanization, agriculture, timber harvest, and road construction destroy, alter, and fragment carnivore habitats. Some predator species have adapted quite well to developed landscapes, and have expanded their populations in proximity to urban and suburban areas. These "urban carnivores," such as raccoons, coyotes, skunks, and raptors, frequently interact with humans. Many of these interactions can be positive for people whose opportunities to view wildlife, particularly carnivores, might otherwise be limited. However, the potential for conflict exists as well, particularly when urban carnivores consume garbage, harass pets, or prey on songbirds or other favored wildlife species. Urban areas also present direct negative impacts on carnivores, including collisions with cars and buildings and increased opportunities for disease transmission.

While urbanization has benefited some carnivores, other species have proven much more sensitive to development and human presence. Urbanization and agricultural practices that radically alter the landscape have excluded many carnivores from large portions of their former ranges. Sprawl, road building, and other activities threaten to fragment and degrade much of what habitat remains. Fragmentation decreases the likelihood of successful dispersal and breeding for carnivores that require large home ranges, such as grizzly bears and Florida panthers. Human activity also brings nonnative species and opens up niches that allow those species to survive, potentially harming native predators through competition, disease transmission, or further alteration of habitats.

The five chapters in Part 2 address human-carnivore coexistence in urban landscapes where human influence is more pronounced. Chapters 4 and 5 discuss predator species that are thriving in urbanized and suburbanized landscapes, and the benefits and conflicts that arise—for both humans and carnivores. Chapter 6 describes how proper design of transportation corridors can reduce both the impacts of habitat fragmentation and the opportunities for negative interaction between people and predators. Chapters 7 and 8 discuss efforts to conserve declining carnivore species in the face of significant, ongoing habitat alteration due to human activities.

Chapter 4, "Ecology and Management of Striped Skunks, Raccoons, and Coyotes in Urban Landscapes," by Stanley D. Gehrt, examines human interactions with species that have adapted successfully to life in urbanized habitats. This chapter discusses the urban ecology of these three medium-sized carnivore species and the ways they have partitioned niches within a 1,500 ha forest preserve northwest of Chicago, and examines how these ecological characteristics influence the relative propensity of each species to come into conflict with humans.

Chapter 5, "Birds of Prey in Urban Landscapes," by R. William Mannan and Clint W. Boal, turns our attention to avian predators and their interactions with humans. This chapter reviews why some species of raptors utilize urban landscapes as habitats; describes the benefits and problems that humans and raptors pose for one another in urbanized areas; and suggests management and education solutions that will help minimize the problems and maximize the benefits.

Chapter 6, "Challenges in Conservation of the Endangered San Joaquin Kit Fox," by Howard O. Clark Jr. et al., examines conservation of a rare canid in rapidly growing California, where population increase and agriculture have radically altered the landscape upon which the San Joaquin kit fox depends. In this instance, the habitat pressures faced by the fox in this changing landscape are exacerbated by another anthropogenic impact:

the introduction of nonnative red foxes, which can competitively exclude the smaller kit fox from what remains of its prime denning and feeding habitat.

Chapter 7, "Carnivore Conservation and Highways: Understanding the Relationships, Problems, and Solutions," by Bill Ruediger, identifies the impacts of development on species that are more sensitive to human activities and developed landscapes. Ruediger discusses how in many parts of the country, roads have reduced and fragmented habitats needed by wide-ranging carnivores. Roads are also a source of car-wildlife collisions. Both problems can be ameliorated with effective wildlife crossings. The chapter discusses the ecological, design, and cost considerations that can make the difference between a marginal and an effective wildlife crossing.

Chapter 8, "Living with Fierce Creatures? An Overview and Models of Mammalian Carnivore Conservation," by David J. Mattson, ties together many of the issues and themes of Parts 1 and 2 and sets the stage for the issues raised in Part 3. Mattson discusses the human-influenced and biological factors governing carnivore conservation, and presents models based on these two categories of causes of endangerment.

Part 2 thus presents a diverse, but by no means comprehensive, selection of the challenges and opportunities faced by carnivores and humans in developed landscapes. As demonstrated here, improvements in science and technology, education, and commitment to conservation can improve the likelihood that humans and carnivores will be able to coexist in these landscapes.

CHAPTER 4

Ecology and Management of Striped Skunks, Raccoons, and Coyotes in Urban Landscapes

Stanley D. Gehrt

The history of carnivores and humans in North America has not been one of amicability or tolerance. Indeed, since European colonization of the continent, conflicts real or perceived have defined the relationship between many species and people. To some extent, this continues today within many urban landscapes. Elsewhere in this volume are discussions regarding relationships between large mammalian carnivores and people within natural, or rural, systems. In this chapter I will characterize the relationships between people and medium-sized carnivores that seem to be highly successful in urban systems: raccoons (*Procyon lotor*), striped skunks (*Mephitis mephitis*), and coyotes (*Canis latrans*). I will explore what is known about the urban ecology of each species and how they become so successful in human-dominated landscapes in North America. The urban ecology for some carnivores is relatively well known, but such is not the case for each of these species. I will focus on ecological similarities and dissimilarities between the species, including the nature of their conflicts with the urban populace.

Conflicts Involving Urban Carnivores

The success of each of these species within urban landscapes provides benefits and costs from a human perspective. As might be predicted, conflicts often occur between these species and people in urban areas, although the nature of these conflicts may differ between species.

Nuisance Issues

Raccoons and skunks frequently run afoul of people in that they may cause property damage and create noise (raccoons) or odors (skunks). Both species may exploit garbage, and thus may pose an inconvenience by creating a mess. In the Chicago metropolitan area, 26% of homeowners considered raccoons to be the species causing the greatest problems in area of residence, second only to Canada geese (Miller et al. 2001). Skunks were considered a problem by 11% of the homeowners; only 2% considered coyotes a nuisance (Miller et al. 2001).

Patterns of nuisance complaints in the Chicago area are probably typical of many metropolitan areas in the United States. According to annual reports submitted by wildlife operators to the Illinois Department of Natural Resources, the number of nuisance raccoons and coyotes captured and handled in the Chicago region increased dramatically from 1991 to 1999, accounting for one-third of the 54,172 nuisance animals handled in 1999 (Bluett et al. 2003). Figure 4.1 illustrates the trend. Gehrt (2003) estimated the economic cost of raccoons to be over $1 million in that year.

Although coyotes were captured less frequently than other species, this species registered the largest relative increase during the 1990s. Coyotes were rare in the Chicago area during most of the twentieth century (Chris Anchor, Forest Preserve District of Cook County, pers. comm.), and rural

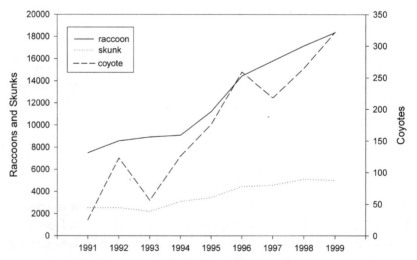

FIGURE 4.1   Numbers of animals handled by wildlife control operators in the Chicago metropolitan area, northeastern Illinois (Bluett et al. 2003).

4. Ecology and Management of Striped Skunks, Raccoons, and Coyotes

areas with hunting and trapping activities were still common in many parts of Cook County through the early to mid-1900s. This was reflected by the low number of coyotes captured in 1991, but between 1991 and 1999 the number of nuisance animals increased from 25 to 322, an increase of 1,188% (see Figure 4.1).

The increase in numbers of nuisance animals trapped each year, particularly for raccoons and coyotes, may be a result of continuing development and urban sprawl, but it also may be a result of increasing wildlife populations within the metropolitan area, as population densities in urban fragments can reach high levels (see below).

Public Health Issues

Raccoons and skunks in urbanized areas represent reservoirs of diseases (Bigler et al. 1975; Rosatte et al. 1991) and parasites (Kazacos 1982) that may affect humans and domestic animals, as well as other species. One of the most important zoonoses (diseases transmittable from animals to humans) worldwide is rabies, and all three species are potential hosts of the virus. Raccoon rabies swept across the Atlantic Coast during the 1980s and 1990s (Beck et al. 1987; Fischman et al. 1992; Riley et al. 1998) and represents one of the most important wildlife diseases in the United States. The disease currently may be spreading west (Uhaa et al. 1992), although extensive control measures may have prevented the epizootic (disease outbreak in animals, equivalent to an "epidemic" in humans) from spreading for the time being. Striped skunks are the primary terrestrial host for rabies over much of the Midwest (Rosatte 1987; Charlton et al. 1991; Greenwood et al. 1997), and epizootics associated with high host densities periodically occur (Verts 1967). Coyotes are the host of a specific strain of rabies that is enzootic (naturally present disease, equivalent to "endemic" in humans) in south Texas, and for the time being it appears to be confined to that region. Prevalence of zoonoses in urban coyote populations has rarely been assessed (Grinder and Krausman 2001a).

Leptospirosis is another enzootic disease in many raccoon and skunk populations (Mitchell et al. 1999), and raccoons have been implicated in human outbreaks of *Leptospirosis interrogans* (Jackson et al. 1993); *L. pomona* (Galton 1959); and *L. autumnalis* (Galton et al. 1959). Additional pathogens frequently associated with raccoons that have important implications for people and domestic animals include pseudorabies (Wright and Thawley 1980); toxoplasmosis (Dubey et al. 1992; Hill et al. 1998;

Mitchell et al. 1999); and, among macroparasites, the raccoon roundworm *Baylisascaris procyonis* (Kazacos and Boyce 1989).

Coyotes present an additional public health issue. Coyotes rarely cause urban property damage, but unlike raccoons and skunks, they occasionally attack and injure or kill housecats and dogs. Rarely, such attacks may extend to people, but these have almost exclusively involved small children. Recent attacks have been reported from Vancouver, B.C., the Boston area, and various municipalities in southern California and Arizona (Grinder and Krausman 1998). Perhaps because of press reports of these incidents, or simply because of their status as the largest carnivores in most municipalities, coyotes are unique in that they are often considered a nuisance prior to causing damage. Indeed, homeowners in the Chicago metropolitan area ranked coyotes as the wildlife species perceived as the greatest threat to human health and safety, with skunks ranked second (Miller et al. 2001). Consequently, simple sightings or the sounds of howling may provide the stimulus for beginning removal programs in some communities.

Benefits of Urban Carnivores

So far I have reviewed the negative aspects of the relationships between people and skunks, raccoons, and coyotes; however, each species may provide important benefits for humans and ecosystem processes. Because of their widespread distribution and easily recognized features, raccoons and skunks are frequently the most common contact many urban residents have with mammalian wildlife species, especially carnivores. Although the sight or sound of a coyote elicits fear in some residents, it often produces more favorable responses in others. The importance of these experiences is difficult to quantify but should not be underestimated. Beyond the aesthetic benefits of such experiences may lie important practical benefits as well. As the proportion of the human population living in urban areas continues to increase, urbanites will have an increasingly important influence on conservation efforts. The importance of conservation issues to urban dwellers is based largely on their perceptions of nature at the local (i.e., urban) level (Middleton 1994). Therefore, experiences urbanites have with urban wildlife, including mammalian carnivores, may have profound impacts on conservation at the regional or national levels (Hadidian et al. 1997; Clergeau et al. 2001).

In addition, each of these species has an ecological role in ecosystem processes, including their role as predators, which are especially poorly understood in urban systems. For example, recent studies in the Chicago metropolitan area suggest coyotes are important predators of white-tailed deer fawns (*Odocoileus virginianus;* Piccolo 2002), and of adult and nestling Canada geese (*Branta canadensis;* C. Paine, Max McGraw Wildlife Foundation, unpublished data). Deer and Canada geese tend to become overpopulated in urban areas, and are often identified as important management issues in urban areas. Therefore coyotes may perform an important practical ecological function through their role as a mortality factor for these species. Indeed, in some urban systems coyotes may be the most important biological control for these species. Coyotes may also benefit some species, such as songbirds, through their control of free-ranging domestic cats (Crooks and Soulé. 1999). However, more research on the roles these predators play in urban systems is clearly needed.

Urban Ecology of Striped Skunks, Raccoons, and Coyotes

It is often assumed that these three species respond to urbanization in similar ways because they presumably occupy similar ecological niches. They each fall into the category mesocarnivore (1–15 kg; *sensu* Buskirk 1999), and each has relatively omnivorous food habits and general habitat requirements. Here I address some ecological attributes of each species and make some interspecific comparisons with regard to the effects of urbanization on their populations.

In addition to summarizing existing literature, I will at times refer to research my colleagues and I have been conducting in the Chicago metropolitan area. In many ways Chicago serves as an excellent model because each of the three species occurs there, and with a human population exceeding 8 million, the landscape is one of the most heavily urbanized areas in North America.

Study Area

We have monitored populations of each of these species in various study areas; however, I will focus on data we have collected at the Ned Brown Forest Preserve (NBFP), a 1,499 ha island of habitat surrounded by a sea of urban development (Figure 4.2). The preserve is located 30 km northwest of Chicago, and is immediately surrounded by substantial residential and industrial development. It is bordered on two sides by major interstates,

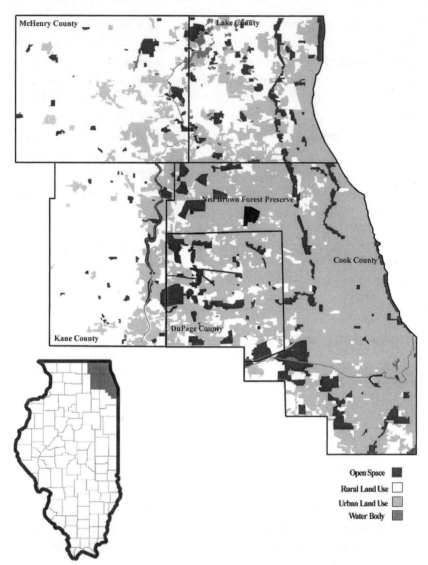

FIGURE 4.2 Location of the Ned Brown Forest Preserve (NBFP) in the Chicago, Illinois, metropolitan area.

and traffic volume is heavy on all adjacent roads, with an average of 302,550 vehicles passing by on adjacent roads every 24 hours (1994 data, Illinois Department of Transportation; Prange et al., in press). Considerable human activity also occurs within the preserve, as it typically receives over 1.5 million visitors annually (Dwyer et al. 1985). Predominant habitats within the preserve include oak woodland, grasslands, and a lake with smaller wetlands (Prange et al. 2003). An important ecological feature is 32 picnic groves distributed within the preserve and containing trash cans and, in some cases, dumpsters. As is typical of all Cook County forest preserves, all trash cans lack lids, dumpsters are left open, and people are excluded from the preserves at night. Consequently, at sundown, ownership of picnic groves essentially transfers from people to nonhuman residents of the preserves.

Monitoring Methods

The raccoon population within a portion of the preserve was monitored using live trapping and radiotelemetry during 1995–2002. Various aspects and phases of this research have been published elsewhere (Gehrt 2002; Prange et al. 2003; Prange et al., in press). During 1995–2001 my colleagues and I captured 647 raccoons, of which a subsample of 90 individuals was radio collared and tracked during diurnal and nocturnal periods. This effort yielded 10,673 radio locations. Similarly, striped skunks have been monitored since 1999 and coyotes since 2000, with both projects still in progress. Skunks were captured throughout the park using livetraps and spotlighting, and coyotes were live-trapped in primarily two small areas (north and south) within the preserve with restricted human access. Through 2002 I captured and radio collared 89 skunks and 11 coyotes at NBFP, with 4,279 locations recorded for skunks and 1,118 locations for coyotes.

Population Density

Comparing density estimates between studies is difficult because researchers have used different methods to estimate population size or conducted sampling efforts during different seasons. Also, densities are often estimated for a study area or trapping grid without consideration for movements outside the study area, which is particularly important, as actual densities will likely be less than estimated densities that use artificial or political boundaries (Prange et al. 2003).

Despite these limitations, comparisons between studies clearly indicate raccoon populations typically occur at much higher densities in urban landscapes than in comparable rural areas. Published rural densities derived from mark-recapture have ranged from 1 to 27 raccoons/km^2 (Moore and Kennedy 1985; Kennedy et al. 1986; Gehrt 1988; Seidensticker et al.

1988). Density estimates reported for raccoons in urbanized areas include 66.7 raccoons/km^2 in a suburban area in Cincinnati (Hoffmann and Gottschang 1977); 111/km^2 in Cincinnati (Schinner and Cauley 1974); and 55.6/km^2 in Toronto (Rosatte et al. 1991). Riley et al. (1998) estimated raccoon densities of 67–333/km^2, and a mean of 125/km^2 for a population in an urban park in Washington, D.C. These are the highest estimates reported from mark-recapture data, although most of their estimates are undoubtedly inflated because the trapping areas were small portions of the park where raccoons concentrated their activity. Raccoon densities also vary by habitat within urban landscapes. In Toronto, the highest population occurred in woodland parks and residential areas, whereas industrial areas and grassland habitats had relatively few raccoons (Rosatte et al. 1991).

In the Chicago study, we used radiotelemetry and standardized trapping designs to estimate the effective trapping area for raccoons, and compared densities between raccoon populations exposed to different levels of anthropogenic effects (Gehrt 2002; Prange et al. 2003). During 1995–2002, the ecological density of the raccoon population at NBFP was 4–12 times greater than a population in a rural park without picnic groves and heavy human visitation (Table 4.1). Spring densities were relatively stable, ranging between 40 and 60 raccoons/km^2, whereas autumn densities were more variable, ranging between 60 and 90 raccoons/km^2. The greater annual variation during autumn may reflect true variation in population size but undoubtedly is also affected by reduced trapability among raccoons during that season (Gehrt and Fritzell 1996).

TABLE 4.1

Mean densities (no. per km^2) and ranges for skunk and raccoon populations at an urban (Ned Brown Forest Preserve, Cook County) and rural (Glacial Park, McHenry County) site in northeastern Illinois. Skunk estimates were obtained during spring and autumn 1991–2001, and raccoon estimates during 1995–1999. Study designs were nearly identical between areas, but different between species.

	Seasons	Urban Site		Rural Site	
		Mean	Range	Mean	Range
Skunk	6	0.86	0.6–1.3	0.76	0.5–1.3
Raccoon	9	47.0	25–87	11.3	1–28

Source: Prange et al. 2003; Gehrt, unpublished data.

Skunk densities in urban areas are less well documented than those for raccoons. To date, there have not been dramatically high densities reported for skunks in urban areas, suggesting that skunks may not be capable of responding to artificial resources and other urban effects to the same degree that raccoons can. Rural estimates from mark-recapture studies include 13–26/km^2 (Ferris and Andrews 1967) and 4–14/km^2 (Verts 1967) in Illinois farmland; 21/km^2 in Manitoba (Lynch 1972); and 4.5/km^2 in Ohio (Bailey 1971). These rural estimates are within the range of estimates from the Toronto metropolitan area, in which skunk densities averaged across the landscape were 2.1–6.5/km^2 but ranged from 1 to 36/km^2 within individual trapping grid cells (Rosatte et al. 1991, 1992).

Our mark-recapture estimate for skunks was 2.1–5.9 skunks/km^2 in 2000. These density estimates were lower than those reported for rural Illinois (Verts 1967), but similar to estimates (2.5–5.5 skunks/km^2) for a population in a rural park using identical protocols (Gehrt, unpublished data). However, traditional mark-recapture estimates from livetrapping were not possible during much of the study because of relatively low trapping success. Therefore, I supplemented my capture protocol with spotlighting and capture by hand, and also expanded my capture effort to the entire forest preserve. This is in stark contrast to the raccoon research, where we had to restrict the livetrapping to a portion of the forest preserve because of the high-density population. Using trapping and spotlighting data to create minimum-known-alive metrics and data from roadkill surveys also indicated similar densities between urban and rural skunk populations (see Table 4.1; Gehrt, unpublished data).

Population density is difficult to estimate for coyotes in any landscape, but particularly so in urban areas, where trapping is usually severely restricted. To date, there have been few indications that coyote densities are higher in urban areas than in similar rural areas, although density estimates have rarely been reported for urban populations. In the Chicago area, preliminary data suggest coyote densities vary widely across the landscape, with highest densities in large urban forest preserves and lower densities in more heavily developed areas. During 2000, I estimated prewhelping coyote densities in three urban parks to range between 0.6 and 3.1 coyotes/km^2 (Gehrt, unpublished data), including an estimate of 1.1 coyotes/km^2 for NBFP. The preserve has had at least two packs residing exclusively within the park throughout the study, one of which had a litter of at least 9 pups in 2002. My estimates were based on livetrapping and visual sightings to provide minimum population estimates, and radiotelemetry data were used to determine the effective area (Gehrt, unpublished data).

Because not all coyotes using an area could be visually sighted, my estimates are undoubtedly conservative. Nevertheless, they are generally higher than typical estimates reported for rural populations, which have typically ranged between 0.2 and 0.5 coyotes/km^2 (Clark 1972; Knowlton 1972; Camenzind 1978; Bowen 1982), with high densities between 0.9 and 2/km^2 (Andelt 1985; Windberg 1995). However, my estimates for parks probably represent the high end of density estimates for urban landscapes, with lower densities likely occurring in other areas lacking natural habitat.

Diet

Diets of carnivores are usually determined from stomach contents or fecal analyses. Stomach content analysis is limited because the animal must be dead, and it only provides information on the diet for a short period prior to death. For these reasons, most information on diets has come from fecal samples. However, fecal analysis is affected by different rates of digestibility for various food items. This issue of digestibility is particularly problematic for urban studies if a species tends to use anthropogenic refuse such as hamburgers or donuts, which are not easily detected in feces. For this reason, we have not conducted traditional diet analyses for raccoons and skunks in the Chicago study, but rather have relied on radiotelemetry data and visual observations.

Raccoons arguably have the most diverse diets of any carnivore (Gehrt, 2003), and this has been important in their success in urban areas. Raccoons readily exploit refuse and other resources related to human activities. The body size of raccoons and their climbing abilities allow them to access trash cans and dumpsters. We typically observed raccoons in picnic groves at NBFP during nighttime tracking, and their exploitation of these resources affected their foraging and spatial pattern (see below). Often raccoons would spend most of the night in and around garbage cans, and feeding aggregations were common (Prange et al., in press).

Skunks are also highly omnivorous (Verts 1967; Rosatte 1987), although the extent to which they exploit refuse and anthropogenic-related resources is poorly understood. It is often assumed that skunks exploit refuse in urban systems, and skunks may exploit residential trash (Rosatte 1987), but I have found few published reports of this. At NBFP, refuse is often present on the ground near trash receptacles, at times in substantial quantities. However, during many nights of radio tracking we rarely observed skunks exploiting trash at picnic groves. Instead, they often foraged in the grassy areas associated with picnic groves, pursuing mice during early spring or insects during summer and autumn. Skunks may have failed to become habitual foragers of refuse (unlike raccoons) because (1) their smaller body size precluded access to trash cans (therefore resources may have been less predictable for skunks than for raccoons); (2) they were outcompeted for resources by raccoons; or (3) they are less socially tolerant than raccoons (which precluded feeding aggregations at picnic groves).

Although coyotes are relatively omnivorous for canids, their diets are comprised primarily of prey. Some coyotes may consume refuse, but we rarely observed coyotes at picnic groves during nocturnal tracking sessions at NBFP. Instead, we observed them foraging in fields or along lake edges. We frequently observed the same behavior outside the preserve in residential, commercial, and park areas (Gehrt, unpublished data). However, we did observe coyotes occasionally feeding on carcasses from vehicle collisions. Roadkill may represent "anthropogenic" resources that differ from typical refuse in that it is typical prey and is not predictable in space or time. Analyses of 325 scats collected in NBFP during 2001 and 2002 yielded usual prey items, such as voles, rabbits, and other mammals (P. Morey, unpublished data). Unfortunately, it is not possible to discriminate between prey scavenged from vehicle collisions and actual predation.

Although coyotes at NBFP and other areas of the Chicago metropolitan area were rarely observed consuming refuse, it is likely some individuals did so. Unfortunately there are few published reports of coyote diets in urban landscapes. Studies of coyote food habits in semirural or suburban areas in California and Arizona have indicated coyotes in those studies did consume refuse and other anthropogenic-related food to varying degrees (MacCracken 1982; Shargo 1988; McClure et al. 1995). Comparisons between these studies are difficult because of different methodologies, but "garbage" constituted 16.7–40.9% of food items in two studies (MacCracken 1982; Shargo 1988).

One aspect of coyote diets that differs from smaller urban carnivores is the occasional consumption of domestic pets, especially feral cats and housecats (*Felis catus*). Domestic cat appeared in 13.6% of coyote scats collected in Malibu, California (Shargo 1988), and it appears occasionally in coyote scats (1.2%) in NBFP (P. Morey, unpublished data). Few cats have been caught or observed in NBFP during my research, although free-ranging housecats are quite common in adjacent residential areas. Negative correlations in spatial distribution between coyotes and cats have been documented for urban landscapes (Crooks and Soulé 1999), with the inference that coyotes remove cats, or cats avoid areas with coyotes.

Movements and Spatial Organization

Comparisons of home range estimates among studies are affected by many of the same factors that affect density estimates: differences in monitoring protocols and home range models, and seasonal differences in movement patterns. Nevertheless, as with patterns of density estimates, it is evident that interspecific patterns of home range variation in urban systems possibly differ between species.

For raccoons, home ranges in urban areas are relatively small in size, which is predictable given that both population size and home range size are influenced by the distribution and quality of resources. Most estimates of home range sizes from a variety of systems and studies are from 50 to 300 ha (see Gehrt, 2003), and home ranges of urban raccoons typically fall within the 5–79 ha range (Schinner and Cauley 1974; Hoffmann and Gottschang 1977; Slate 1985; Rosatte et al. 1991; Feigley 1992).

At NBFP, female raccoons were intensively monitored in each season during 1996–1997 (Prange et al., in press), and median seasonal home range sizes ranged between 25.2 and 52.8 ha, with some home ranges as small as 11 ha. These home ranges were significantly smaller than estimates from a rural population that was monitored simultaneously (Prange et al., in press). Table 4.2 presents average home range estimates for both

TABLE 4.2

Mean (SD) seasonal home range sizes (ha) for radiocollared raccoons during 1998–2001 at Ned Brown Forest Preserve, Chicago. Home ranges were estimated with the 95% adaptive kernel (AK) model.

Gender	Season	Skunk		Raccoon	
		Home Range	n	Home Range	n
Female	Summer	168 ±75	17	109 ±88	17
Male	Summer	98 ±1	2	128 ±67	24
Female	Winter	101 ±89	16	77 ±42	13
Male	Winter	137 ±103	5	102 ±52	30

sexes of raccoons at NBFP during 1998–2001, which are similar to lower home range estimates from other studies.

The distribution and quality of artificial resources at NBFP, represented by picnic groves, had a profound effect on the spatial pattern of raccoon home ranges. Many raccoons focused their nocturnal activity near picnic areas (Figure 4.3), and raccoon home ranges at NBFP were more highly aggregated and more stable (with regard to seasonal variation in size and shifts in activity centers) than in rural populations with more evanescent resources (Prange et al., in press). It was typical during nocturnal radio-tracking sessions to observe groups of 3–7 raccoons foraging in close proximity at a single picnic area. Therefore, it appears that use of these resources was not temporally partitioned, and that urban raccoons were relatively socially tolerant (Prange et al., in press).

For skunks, differences in home range size between urban and rural populations are less clear than for raccoons. Reported home ranges of

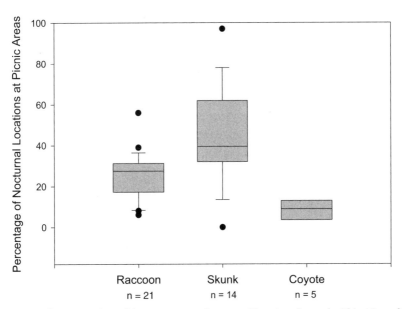

FIGURE 4.3 Box plots of the percentage of nocturnal locations located within 15 m of a picnic grove for an individual during summer at the Ned Brown Forest Preserve, Chicago, Illinois. Only individuals with >30 nocturnal locations during the season are included, and data for individuals monitored during multiple years were averaged across years. Data for raccoons were recorded during 1995–1999; for skunks during 1999–2001; and for coyotes during 2000–2001.

skunks in rural areas have typically varied between 30 ha and 500 ha (Storm 1972; Rosatte and Gunson 1984; Greenwood et al. 1997). Average home range size was 270 ha in Alberta (no difference between sexes; Rosatte and Gunson 1984) and in Saskatchewan was 1,210 ha for males and 430 ha for females (Lariviere and Messier 1998). Home range size in Toronto averaged 64 ha (Rosatte et al. 1991), which is considerably less than typical rural estimates, and suggests that skunk home range size decreases in urban areas. However, seasonal skunk home ranges at NBFP averaged 98–168 ha (see Table 4.1), which was not significantly different from home ranges for a rural population monitored simultaneously in Illinois (Gehrt, unpublished data) and from some published estimates (Storm 1972). It is important to note that many rural skunk studies reporting home ranges have been conducted on the northern Great Plains, and may not be a valid comparison with many urban systems in other parts of the country.

Interestingly, average home range sizes for skunks at NBFP were similar to those for raccoons (see Table 4.2), despite the larger size of raccoons. This is likely the result of different foraging patterns for the two species. Like raccoon home ranges, skunk home ranges were aggregated with varying degrees of spatial overlap, but unlike raccoon home ranges, they were not necessarily centered around picnic groves but rather around denning areas (Gehrt, unpublished data). Many skunks foraged near picnic areas most of the time, and the mean percentage of locations in picnic areas among skunks was actually higher than the mean for raccoons (see Figure 4.3). However, it became apparent during 229 nights of monitoring and resulting visual observations that they were rarely exploiting refuse but rather taking advantage of the prey associated with the mowed lawns in these areas. We have not yet observed foraging aggregations among skunks.

Published home ranges for coyotes have varied greatly among studies, and among individuals within a population. Coyote populations differ from skunks and raccoons in that they are typically comprised of territorial individuals or packs and solitary, transient individuals that have relatively large home ranges (Andelt 1985). Thus, differences in home range size between urban and rural populations have been equivocal, although few data exist for major metropolitan areas. Composite home range size in urban Los Angeles averaged 1.1 km^2 (Shargo 1988), which was similar to the lower end of the range reported for resident coyotes in Tucson (Grinder and Krausman 2001b), although this range was considerable (1.7–59.7

km^2). Most published home ranges of resident coyotes in rural or undisturbed systems average 10–31 km^2 (Bekoff and Wells 1980; Litvaitis and Shaw 1980; Bowen 1982; Roy and Dorrance 1985; Gese et al. 1988; Servin and Huxley 1995). In a high-density population in rural Texas, average home range size was only 4–5 km^2 for residents (Andelt 1985).

Preliminary data from the Chicago metropolitan area yielded an average home range size of 9 km^2 for territorial packs (P. Morey and S. D. Gehrt, unpublished data), which is similar to 12.5–13.2 km^2 averages reported for residents in Tucson (Grinder and Krausman 2001b). However, average Chicago home ranges are much smaller than comparable estimates (medians 16.8–27.0 km^2) for coyotes in an agricultural landscape in Illinois (Gosselink et al. 2003). Transients with unusually large home ranges were reported for Tucson (Grinder and Krausman 2001b) and have also been documented in Chicago (Gehrt, unpublished data).

Radio-collared coyotes in NBFP did not center their foraging at picnic areas (see Figure 4.3), which is consistent with fecal analyses from the area (Morey and Gehrt, unpublished data). Coyote radio locations were recorded in all areas of the preserve. Occasionally, individual coyotes were briefly seen at picnic areas, so they may have taken advantage of refuse when they encountered it, but unlike raccoon foraging, their foraging apparently was not heavily influenced by artificial resources. However, we do not understand the influence of roadkill on coyote movement patterns. It is possible that roadkill in heavily urbanized areas is frequent enough that coyotes may forage along roads.

The spatial patterns of coyote locations in NBFP were typical of other large (>450 ha) urban preserves in the Chicago area. Unlike most raccoons and skunks, coyote movements encompassed the forest preserve, and the size and shape of the home range of residents closely followed the boundaries of the forest preserve. This may explain why urban studies report a considerable range in home range size of residents. Thus, for coyotes residing in large urban natural areas, anthropogenic resources have less of an effect on home range shape or size than do development and anthropogenic activities adjacent to the area.

Anthropogenic effects on the spatial pattern and movements of the three species differ on a large spatiotemporal scale, as indicated by Figure 4.4. Although relatively more raccoons have been monitored for a longer period than skunks or coyotes, the area encompassed by racoons' cumulative locations is smaller than that of the other two species. Few raccoons or

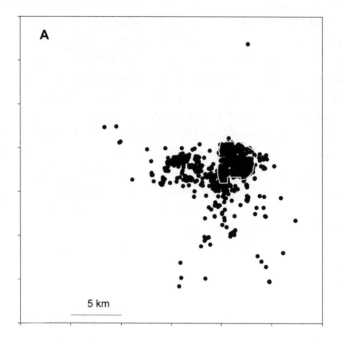

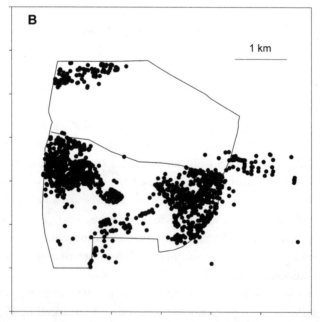

FIGURE 4.4 Spatial distribution of cumulative radiolocations for (A) coyotes, (B) striped skunks, and (C) *(facing page)* raccoons captured at Ned Brown Forest Preserve (NBFP) in the Chicago, Illinois, metropolitan area during 1995–2001.

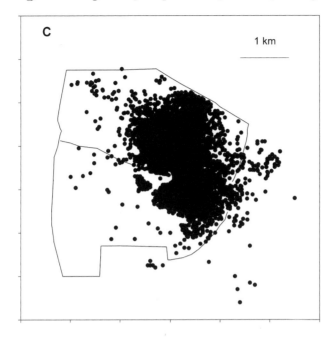

skunks regularly left the preserve (but see Prange et al., in press), but most coyotes (particularly pups and subadults) eventually left. No dispersal from NBFP was documented for eartagged or radio-collared raccoons during the study, and no raccoons were documented crossing interstates to the west or north of the preserve. Similarly, emigration from NBFP by radio-collared striped skunks has been a rare event. Although resident coyotes rarely venture from the preserve (Figure 4.4), surrounding development and traffic apparently did not inhibit immigration or emigration.

A Note on Cultural Inheritance

Behavioral plasticity is an important characteristic for adapting to urban landscapes, and learned behavior may be passed on to future generations in some of these species. However, this arena has received little attention from researchers, and much of what follows is relatively speculative. Cultural inheritance may occur in raccoons (Gehrt, 2003) and may help explain why some raccoons are more "urbanized" than others. Hadidian et al. (1991) described discreet choices among individual raccoons for types of diurnal rest sites in an urban area. Some raccoons exclusively selected buildings for rest sites, whereas others restricted their use to "natural" dens within a park. Similarly, raccoons at NBFP almost exclusively used tree cavities as rest sites, despite the apparent availability of many buildings nearby. Yet, some raccoons adjacent to NBFP apparently used buildings as rest sites (as indicated by nuisance reports), which suggests that within urban landscapes individual raccoons may differ in their preference for anthropogenic resources, including den sites. Young raccoons may maintain familial relationships for nearly a year until their mother comes into estrus (Gehrt and Fritzell 1998) and probably develop search images during that period that are influenced by the mother.

Cultural inheritance may also play a role in the urbanization of coyotes. Most research of coyote habitat use in rural midwestern landscapes has described coyotes as avoiding buildings and other developments associated with human activities (Gosselink et al. 2003). Undoubtedly this pattern of habitat use is related to coyotes associating humans with a threat, as sport hunting and harvest are the most common sources of mortality for coyotes in rural areas (Gosselink et al. 2003). In urban systems, where sport hunting and trapping are essentially nonexistent, generations of coyotes are reared in close proximity to humans, and a different type of land use pattern may develop in those coyotes. Extended familial relationships beyond weaning are typical in coyotes, providing the opportunity for parents to influence the behavior of their young.

In contrast to familial relationships in raccoons and coyotes, familial relationships for striped skunks are relatively brief, with juveniles typically becoming independent at 3 months of age (Verts 1967; Gehrt, unpublished data). Thus, there appear to be limited opportunities for cultural transmission in skunks, although the social organization of free-ranging populations is poorly understood. This presents the interesting possibility that skunks are less adaptable to urban systems than are raccoons and coyotes because of relatively rigid behavioral constraints or limited familial contact. Obviously this is speculative but warrants further investigation for these and other species.

Management Implications

Understanding how populations respond to anthropogenic stimuli will assist with developing appropriate management strategies for carnivores that occur in urban areas. Information on the urban ecology of wildlife species

4. Ecology and Management of Striped Skunks, Raccoons, and Coyotes

can have direct management implications. For example, current evidence suggests reduction of trash (or access to trash) in urban parks and residential areas will reduce the carrying capacity for raccoon populations, with resulting reductions in density and possibly reproductive rate (Prange et al. 2003). However, such a manipulation has not yet been published to verify this prediction.

Removing access to refuse may not result in the same response in skunk and coyote populations, although removing trash and discouraging wildlife feeding by residents may affect the behavior of certain individual animals, thereby reducing human-wildlife conflicts. Monitoring of coyotes in the Chicago area has revealed that the majority of coyotes using this area have few negative interactions with people or pets. Nevertheless, discouraging feeding by residents may help reduce habituation by some individual coyotes at the local level. A common theme for most incidents of coyote attacks on people has been the loss of fear by coyotes as a result of consistent access to anthropogenic resources, whether refuse or intentional feeding (Carbyn 1989; Baker and Timm 1998).

Sociospatial patterns have implications for the translocation of nuisance or rehabilitated wildlife. Translocation programs often seek to release nuisance coyotes in large natural patches or refuges, yet it is precisely these areas that are most likely to contain territorial packs. In the Chicago area, radio-collared nuisance coyotes translocated to a large semiurban forest preserve typically dispersed within 24 hours (Gehrt, unpublished data), most likely because the forest preserve was already occupied by a pack of coyotes. However, this type of dispersal pattern is typical not only of territorial species. Nuisance raccoons translocated to a rural forest preserve usually disperse within a few days, and often end up as nuisance animals again (Mosillo et al. 1999). It is unknown whether nuisance raccoons dispersed from translocation sites because resident raccoons forced them or because they were attracted to buildings (via cultural inheritance).

Cultural inheritance may affect other aspects of nuisance management in addition to translocation of raccoons. Baker and Timm (1998) suggested limited removal efforts will be more effective for coyotes that have recently inhabited urban areas or become habituated to human activities. Similarly, we have been able to modify the behavior of a pack with the capture of a single individual, thereby instilling an avoidance of humans and their activities. More extensive trapping efforts and the removal of more individuals are necessary for coyotes with a long history of coexistence with the urban populace (Baker and Timm 1998).

A consistent theme of successful management programs for these species is the importance of public education in reducing potential conflicts. Modifying human behavior to reduce availability of food or den sites in areas unsuitable for wildlife may be a more effective management strategy than managing nuisance wildlife post hoc. Effective management programs for coyotes in southern Californian municipalities have emphasized public education first, with removal or population reduction as secondary alternatives (Baker and Timm 1998). Similarly, discouraging feeding of raccoons by residents will reduce the number of animals associating buildings or people with food, thereby reducing conflicts and possible exposure to disease.

Trapping for removal or translocation of nuisance individuals is usually conducted by private operators contracted by property owners. However, the removal of nuisance individuals can be difficult (particularly with coyotes), and the success of the removal will be temporary unless the reasons for the conflict can be identified and corrected.

Conclusion

The NBFP serves as a microcosm within a larger urban landscape that illustrates differences and similarities in the responses of each species to anthropogenic effects. The relative abundance of each species in the preserve reflects numbers of nuisance animals captured in the area at large, and also relative abundance of raccoons and skunks in metropolitan Toronto (Rosatte et al. 1992). Each of the three species benefited from protection from hunting and trapping, which represent important causes of mortality for rural populations. However, each species differed in their response to artificial resources and adjacent development. In spite of their ecological similarities and apparent success in urban landscapes, striped skunks, raccoons, and coyotes exhibit differences in their response to urbanization and the products of human activities. My review should be considered a preliminary attempt to illuminate the varied pathways to success in urban systems that are followed by different carnivore species.

Acknowledgments

My research, which comprises the foundation for this chapter, was conducted while I was affiliated with the Max McGraw Wildlife Foundation. In addition, the Illinois Department of Natural Resources supported raccoon and skunk research projects through the Furbearer Fund, and the raccoon research was further supported by Critter Control, Inc., and the Rice Foundation. The coyote research was supported by Cook County Animal Control. The Forest Preserve District of Cook County and Chris Anchor provided additional support for each project. I am indebted to graduate students Suzie Prange and Paul Morey, and a large contingent of technicians, for excellent fieldwork in trying conditions. Robert Bluett made editorial suggestions that improved the manuscript.

Literature Cited

Andelt, W. F. 1985. "Behavioral ecology of coyotes in south Texas," *Wildlife Monographs* 94. Wildlife Society, Washington, DC.

Bailey, T. N. 1971. "Biology of striped skunks on a southwestern Lake Erie marsh," *American Midland Naturalist* 85:196–207.

Baker, R. O., and R. M. Timm. 1998. "Management of conflicts between urban coyotes and humans in southern California," in *Proceedings of the Eighteenth Vertebrate Pest Conference*, ed. R. O. Baker and A. C. Crabb, 299–312. University of California, Davis.

Beck, A. B., S. R. Felser, and L. T. Glickman. 1987. "An epizootic of rabies in Maryland, 1982–1984," *American Journal of Public Health* 77:42–44.

Bekoff, M., and M. C. Wells. 1980. "The social ecology of coyotes," *Scientific American* 242:130–148.

Bigler, W. J., J. H. Jenkins, P. M. Cumbie, G. L. Hoff, and E. C. Prather. 1975. "Wildlife and environmental health: Raccoons as indicators of zoonoses and pollutants in southeastern United States," *Journal of the American Veterinary Medical Association* 167:592–597.

Bluett, R. D., G. F. Hubert Jr., and C. A. Miller. 2003. "Regulatory oversight and activities of wildlife control operators in Illinois," *Wildlife Society Bulletin* 31:104–116.

Bowen, W. D. 1982. "Home range and spatial organization of coyotes in Jasper National Park, Alberta," *Journal of Wildlife Management* 46:201–216.

Buskirk, S. W. 1999. "Mesocarnivores of Yellowstone," in *Carnivores in Ecosystems: The Yellowstone Experience*, ed. T. W Clark, P. M. Curlee, S. C. Minta, and P. M. Kareiva, 165–187. Yale University Press, New Haven, CT.

Camenzind, F. J. 1978. "Behavioral ecology of coyotes on the National Elk Refuge, Jackson, Wyoming," in *Coyotes: Biology, Behavior and Management*, ed. M. Bekoff, 267–274. Academic Press, New York.

Carbyn, L. N. 1989. "Coyote attacks on children in western North America," *Wildlife Society Bulletin* 17:444–446.

Charlton, K. M., W. A. Webster, and G. A. Casey. 1991. "Skunk rabies," in *The Natural History of Rabies*, 2nd edition, ed. G. M. Baer, 307–324. CRC Press, Boca Raton, FL.

Clark, F. W. 1972. "Influence of jackrabbit density on coyote population change," *Journal of Wildlife Management* 36:343–356.

Clergeau, P., J. Jokimaki, and J. L. Savard. 2001. "Are urban bird communities influenced by the bird diversity of adjacent landscapes?" *Journal of Applied Ecology* 38:1122–1134.

Crooks, K. R., and M. E. Soulé. 1999. "Mesopredator release and avifaunal extinctions in a fragmented system," *Nature* 400:563–566.

Dubey, J. P., A. N. Hamir, C. A. Hanlon, and C. E. Rupprecht. 1992. "Prevalence of *Toxoplasma gondii* infection in raccoons," *Journal of the American Veterinary Medical Association* 200:534-536.

Dwyer, J. F., H. W. Schroeder, and R. L. Buck. 1985. "Patterns of use in an urban forest recreation area," in *Proceedings of the 1985 National Outdoor Recreation Trends Symposium, vol. 2*, ed. J. Wood, 81-89. U. S. Department of the Interior, Atlanta, GA.

Feigley, H. P. 1992. "The ecology of the raccoon in suburban Long Island, N.Y., and its relation to soil contamination with *Baylisascaris procyonis ova*." Ph.D. dissertation, State University of New York, Syracuse.

Ferris, D. H., and R. D. Andrews. 1967. "Parameters of a natural focus of *Leptospira pomona* in skunks and opossums," *Bulletin of the Wildlife Disease Association* 3:2-10.

Fischman, H. R., J. T. Horman, and E. Israel. 1992. "Epizootic of rabies in raccoons in Maryland," *Journal of the American Veterinary Medical Association* 201:1883-1886.

Galton, M. M. 1959. "The epidemiology of leptospirosis in the United States," *Public Health Report* 74:141-148.

Galton, M. M., N. Hirschberg, R. W. Menges, M. P. Hines, and R. Habermann. 1959. "An investigation of possible wild animal hosts of leptospires in the area of the 'Fort Bragg fever' outbreaks," *American Journal of Public Health* 49:1343-1348.

Gehrt, S. D. 1988. "Movement patterns and related behavior of the raccoon, *Procyon lotor*, in east-central Kansas." M.S. thesis, Emporia State University, Emporia, KS.

———. 2002. "Evaluation of spotlight and road-kill surveys as indicators of local raccoon abundance," *Wildlife Society Bulletin* 30:449-456.

———. 2003. "Raccoons (*Procyon lotor* and Allies)," in *Wild Mammals of North America: Biology, Management and Conservation*, 2nd Edition, ed. G. A. Feldhamer, B. C. Thompson and J. A. Chapman, 611-634. Johns Hopkins University Press, Baltimore, MD.

Gehrt, S. D., and E. K. Fritzell. 1996. "Sex-biased response of raccoons (*Procyon lotor*) to live traps," *American Midland Naturalist* 135:23-32.

———. 1998. "Resource distribution, female home range dispersion and male spatial interactions: Group structure in a solitary carnivore," *Animal Behaviour* 55:1211-1227.

Gese, E. M., O. J. Rongstad, and W. R. Mytton. 1988. "Home range and habitat use of coyotes in southeastern Colorado," *Journal of Wildlife Management* 52:640-646.

Gosselink, T. E., T. R. Van Deelen, R. E. Warner, and M. G. Joselyn. 2003. "Temporal habitat partitioning and spatial use of coyotes and red foxes in east-central Illinois," *Journal of Wildlife Management* 67:90-103.

Greenwood, R. J., W. E. Newton, G. L. Pearson, and G. J. Schamber. 1997. "Population and movement characteristics of radio-collared striped skunks in North Dakota during an epizootic of rabies," *Journal of Wildlife Diseases* 33:226-241.

Grinder, M. I., and P. R. Krausman. 1998. "Ecology and management of coyotes in Tucson, Arizona," *Proceedings of the Vertebrate Pest Conference* 18:293-298.

———. 2001a. "Morbidity-mortality factors and survival of an urban coyote population in Arizona," *Journal of Wildlife Diseases* 37:312-317.

———. 2001b. "Home range, habitat use, and nocturnal activity of coyotes in an urban environment" *Journal of Wildlife Management* 65:887-898.

Hadidian, J., D. A. Manski, and S. Riley. 1991. "Daytime resting site selection in an urban raccoon population," in *Proceedings of the Second National Symposium on Urban Wildlife, Cedar Rapids, Iowa*, ed. L. W. Adams and D. L. Leedy, 39-45. National Institute for Urban Wildlife, Columbia, MD.

Hadidian, J., J. Sauer, C. Swarth, P. Handley, S. Droege, C. Williams, J. Huff, and G. Didden. 1997. "A citywide breeding bird survey for Washington D.C.," *Urban Ecosystems* 1:87-102.

Hill, R. E., J. J. Zimmerman, R. W. Wills, S. Patton, and W. R. Clark. 1998. "Seroprevalence of antibodies against *Toxoplasma gondii* in free-ranging mammals in Iowa," *Journal of Wildlife Diseases* 34:811–815.

Hoffmann, C. O., and J. L. Gottschang. 1977. "Numbers, distribution, and movements of a raccoon population in a suburban residential community," *Journal of Mammalogy* 58:623–636.

Jackson, L. A., A. F. Kaufman, W. G. Adams, M. B. Phelps, C. Andreasen, C. W. Langkop, B. J. Francis, and J. D. Wenger. 1993. "Outbreak of leptospirosis associated with swimming," *Pediatric Infectious Disease Journal* 12:48–54.

Kazacos, K. R. 1982. "Contaminative ability of *Baylisascaris procyonis* infected raccoons in an outbreak of cerebrospinal nematodiasis," *Proceedings of the Helminthological Society of Washington* 49:155–157.

Kazacos, K. R., and W. M. Boyce. 1989. "*Baylisascaris* larva migrans," *Journal of the American Veterinary Medical Association* 195:894–903.

Kennedy, M. L., G. D. Baumgardner, M. E. Cope, F. R. Tabatabai, and O. S. Fuller. 1986. "Raccoon (*Procyon lotor*) density as estimated by the census-assessment line technique," *Journal of Mammalogy* 67:166–168.

Knowlton, F. F. 1972. "Preliminary interpretations of coyote population mechanics with some management implications" *Journal of Wildlife Management* 36:369–382.

Lariviere, S., and F. Messier. 1998. "Spatial organization of a prairie striped skunk population during the waterfowl nesting season," *Journal of Wildlife Management* 62:199–204.

Litvaitis, J. A., and J. H. Shaw. 1980. "Coyote movements, habitat use, and food habits in southwestern Oklahoma," *Journal of Wildlife Management* 44:62–68.

Lynch, G. M. 1972. "Effect of strychnine control on nest predators of dabbling ducks," *Journal of Wildlife Management* 36:436–440.

MacCracken, J. G. 1982. "Coyote foods in a southern California suburb," *Wildlife Society Bulletin* 10:280–281.

McClure, M. F., N. S. Smith, and W. W. Shaw. 1995. "Diets of coyotes near the boundary of Saguaro National Monument and Tucson, Arizona," *Southwestern Naturalist* 40:101–104.

Middleton, J. 1994. "Effects of urbanization on biodiversity in Canada," in *Biodiversity in Canada: A Science Assessment for Environment Canada,* 115–120. Environment Canada, Ottawa.

Miller, C. A., L. K. Campbell, and J. A. Yeagle. 2001. "Attitudes of homeowners in the greater Chicago metropolitan region toward nuisance wildlife," *Human Dimensions Program Report SR-00-02.* Illinois Natural History Survey, Champaign, IL.

Mitchell, M. A., L. L. Hungerford, C. Nixon, T. Esker, J. Sullivan, R. Koerkenmeier, and J. P. Dubey. 1999. "Serologic survey for selected infectious disease agents in raccoons from Illinois," *Journal of Wildlife Diseases* 35:347–355.

Moore, D. W., and M. L. Kennedy. 1985. "Weight changes and population structure of raccoons in western Tennessee," *Journal of Wildlife Management* 49:906–909.

Mosillo, M., E. J. Heske, and J. D. Thompson. 1999. "Survival and movements of translocated raccoons in northcentral Illinois," *Journal of Wildlife Management* 63:278–286.

Piccolo, B. P. 2002. "Behavior and mortality of white-tailed deer neonates in suburban Chicago, Illinois." M.S. thesis, University of Illinois, Urbana.

Prange, S., S. D. Gehrt, and E. P. Wiggers. 2003. "Demographic factors contributing to high raccoon densities in urban landscapes," *Journal of Wildlife Management* 67:324–333.

———. In press. "Influences of anthropogenic resources on raccoon (*Procyon lotor*) movements and spatial distribution," *Journal of Mammalogy.*

Riley, S. P. D., J. Hadidian, and D. A. Manski. 1998. "Population density, survival, and rabies in raccoons in an urban national park," *Canadian Journal of Zoology* 76:1153–1164.

Rosatte, R. C. 1987. "Striped, spotted, hooded, and hog-nosed skunk," in *Wild Furbearer Management and Conservation in North America,* ed. M. Novak, J. Baker, M. Obbard, and B. Malloch, 599-613. Ontario Trappers Association, North Bay, Ontario.

Rosatte, R. C., and J. R. Gunson. 1984. "Dispersal and home range of striped skunks, *Mephitis mephitis,* in an area of population reduction," *Canadian Field-Naturalist* 98:315-319.

Rosatte, R. C., M. J. Power, and C. D. MacInnes. 1991. "Ecology of urban skunks, raccoons, and foxes in metropolitan Toronto," in *Wildlife Conservation in Metropolitan Environments,* ed. L. W. Adams and D. L. Leedy, 31-38. National Institute for Urban Wildlife, Columbia, MD.

———. 1992. "Density, dispersion, movements and habitat of skunks (*Mephitis mephitis*) and raccoons (*Procyon lotor*) in metropolitan Toronto," in *Wildlife 2001: Populations,* ed. D. E. McCullough and R. E. Barrett, 932-944. Elsevier Science Publishers LTD, Essex, UK.

Roy, L. D., and M. J. Dorrance. 1985. "Coyote movements, habitat use, and vulnerability in central Alberta," *Journal of Wildlife Management* 49:307-313.

Schinner, J. R., and D. L. Cauley. 1974. "The ecology of urban raccoons in Cincinnati, Ohio," in *Wildlife in an Urbanizing Environment,* ed. J. H. Noyes and D. R. Progulske, 125-130. Planning and Resource Development Series No. 28, Holdsworth Natural Resources Center, Amherst, MA.

Seidensticker, J., A. J. T. Johnsingh, R. Ross, G. Sanders, and M. B. Webb. 1988. "Raccoons and rabies in Appalachian mountain hollows," *National Geographic Research* 4:359-370.

Servin, J., and C. Huxley. 1995. "Coyote home range size in Durango, Mexico," *Zeitschrift für Säugetierkunde* 60:119-120.

Shargo, E. S. 1988. "Home range, movements, and activity patterns of coyotes (*Canis latrans*) in Los Angeles suburbs." Ph.D. dissertation, University of California, Los Angeles.

Slate, D. 1985. "Movement, activity, and home range patterns among members of a high density suburban raccoon population." Ph.D. dissertation, Rutgers University, New Brunswick, NJ.

Storm, G. L. 1972. "Daytime retreats and movements of skunks on farmland in Illinois," *Journal of Wildlife Management* 36:31-45.

Uhaa, I. J., V. M. Dato, F. E. Sorhage, J. W. Berkley, D. E. Roscoe, R. D. Gorsky, and D. B. Fishein. 1992. "Benefits and costs of using an orally absorbed vaccine to control rabies in raccoons," *Journal of the American Veterinary Medical Association* 201:1873-1882.

Verts, B. J. 1967. *The Biology of the Striped Skunk.* University of Illinois Press, Urbana.

Windberg, L.A. 1995. "Demography of a high-density coyote population," *Canadian Journal of Zoology* 73:942-954.

Wright, J. C., and D. G. Thawley. 1980. "Role of the raccoon in transmission of pseudorabies: A field and laboratory investigation," *American Journal of Veterinary Research* 41:581-583.

CHAPTER 5

Birds of Prey in Urban Landscapes

R. William Mannan and Clint W. Boal

Urban landscapes generally include a gradient of environments with highly developed land, such as the centers of towns and cities, on one end of the continuum, and lightly developed land on the other (Blair 1996). Some species of kites, hawks, falcons, owls, and eagles, collectively called birds of prey, occasionally reside in urban landscapes in close association with people (for a review see Adams 1994; Gehlbach 1994; Smith et al. 1999; Anderson and Plumpton 2000). Birds of prey may occupy urban landscapes as year-round residents, during breeding or nonbreeding seasons only, or for short, intermittent periods (e.g., stopover sites during migration) (Anderson and Plumpton 2000). A few species such as the Mississippi kite (*Ictinia mississippiensis;* Parker 1996); merlin (*Falco columbarius;* Sodhi et al. 1992); Cooper's hawk (*Accipiter cooperii;* Rosenfield et al. 1995; Boal and Mannan 1998); and Eastern screech-owl (*Megascops asio;* Gehlbach 1994) may be more abundant in some towns and cities than in undeveloped areas.

The presence of birds of prey in urban landscapes, particularly at the highly developed end of the urban gradient, has increased the opportunity for predatory birds and people to interact. Outcomes of these interactions can be positive or negative for both people and birds. In this chapter, we briefly review proposed explanations for why birds of prey occupy urban landscapes. We also describe the positive and negative aspects associated with having birds of prey live in close proximity to people, and outline potential ways to reduce problems created by this arrangement.

Why Do Birds of Prey Use Urban Landscapes?

Birds of prey, like other mobile organisms, select places to live that have the resources they need for survival and reproduction (i.e., the places are "habitat;" Morrison et al. 1998). Thus, birds of prey live in urban landscapes because the environmental cues that trigger them to settle, such as cover for nesting or abundant food, are present (Adams 1994; also see general models by Gehlbach 1996; Parker 1996). The suite of environmental cues and associated resources necessary to trigger residency in urban areas likely varies among species. For example, in Tucson, Arizona, red-tailed hawks (*Buteo jamaicensis*) nest in the least-developed environments, whereas Cooper's hawks nest at the other end of the urban gradient in residential areas with high densities of houses (Mannan et al. 2000).

Three scenarios also may explain, in more detail, why some birds of prey identify urban landscapes as habitat (Anderson and Plumpton 2000). First, predatory birds may persist in undeveloped areas that are preserved as parks or natural open space, as urban areas grow around them (e.g., Stewart et al. 1996). Similarly, birds of prey may occupy areas even though some development has taken place, if the changes wrought by development do not alter the environmental cues or resources that trigger residency. For example, a few houses constructed in a woodland area may have little influence on the likelihood of residency by some woodland hawks (e.g., Bloom et al. 1993). In both of these situations, areas of natural landscapes are largely unchanged by urbanization; therefore, the presence of birds of prey in them is not surprising.

A second explanation for why some birds of prey occupy urban landscapes is that exotic plants or human structures in urban areas mimic natural habitat features and, once created, trigger residency by birds of prey, either through natural colonization or reintroduction by people (Anderson and Plumpton 2000). In Tucson, Cooper's hawks most frequently nest in small groves of exotic trees (Boal and Mannan 1998), which apparently mimic the structure of favored nest sites outside of urban areas in the Sonoran desert (e.g., cottonwood [*Populus fremontii*] groves in riparian zones). The exotic trees in Tucson are used by hawks only after they grow for 40 to 50 years; thus, landscaping of private yards and apartment complexes in the uplands of the Sonoran desert has created, over time, nesting habitat for Cooper's hawks where little existed before urbanization. Similarly, trees planted by settlers in the North American prairies, and tall buildings in some cities, apparently mimic the structure of natural nest sites for Mississippi kites (Parker 1996) and peregrine falcons (*Falco peregrinus*) (Cade and Bird 1990), respectively.

A third explanation involves the tendency of individuals of some bird species to try to nest in the same places where they have previously had success, despite modifications in the environment (i.e., "site tenacity") (Wiens 1985). Thus, if a pair of predatory birds successfully nested in an area that subsequently was developed, the pair might renest in that area, despite the changes. If choice of nesting habitat by their offspring was influenced by early experience, a process known as "habitat imprinting" (Wecker 1963), then use of developed areas by members of this species might increase over time (Temple 1977).

The explanations offered above are dependent on the ability of birds to tolerate human activity and disturbance. It is unclear how some birds of prey become acclimated to the presence of humans (Marzluff et al. 1998), but the process may be related, at least in part, to a decrease over the last several decades in the level of active and lethal persecution (i.e., shooting and poisoning) of predatory birds by humans. The change in human behavior toward birds of prey likely is a product of a combination of education, law enforcement, and a shift from rural to urban cultures where people have fewer reasons to be concerned about the presence of predators.

Birds of Prey in Urban Landscapes

The presence of birds of prey in urban landscapes brings benefits and problems for both the birds and humans.

Benefits to Birds of Prey

The most obvious benefit to birds of prey that are able to occupy urban landscapes is, from a population perspective, that a portion of their habitat is maintained, despite being modified by development. In situations where urbanization creates resources for birds of prey (e.g., nest sites), the amount of habitat available to some species may increase as urban landscapes expand (e.g., Mississippi kites; Parker 1996).

Moreover, resources in urban areas may be rich relative to undeveloped areas. Eastern screech-owls nesting in older suburban areas have relatively high nesting success possibly because of greater prey resources (mostly birds) compared with rural areas (Gehlbach 1994, 1996). The high nesting densities attained by some species in urban areas (e.g., Cooper's hawks; Rosenfield et al. 1995) also may be evidence of rich resources. For example, the size of home ranges of male Cooper's hawks breeding in Tucson is small relative to that reported in more natural environments and suggests that the urban hawks do not have to range far to find food (Mannan and Boal 2000). This idea is supported by evidence indicating that urban areas often have a higher total density and biomass of passerine and other small birds, the primary prey for Cooper's hawks, than nonurban areas (e.g., in Tucson, Emlen 1974; and elsewhere, Beissinger and Osborne 1982; Blair 1996; Marzluff et al. 1998). In addition, male Cooper's hawks in Tucson deliver approximately twice the number of prey items and biomass to females and nestlings when compared with males outside of urban areas (Estes and Mannan 2003). Abundance of food associated with human development (i.e., agriculture) also is partly responsible for the presence of nesting Mississippi kites in nearby urban areas (Parker 1996).

Resources for birds of prey in urban areas also may be more stable (Gehlbach 1996). For example, drought in the southwestern United States likely reduces primary productivity of the desert surrounding developed areas, but people ameliorate these effects by watering their lawns and landscaped plants. In contrast, nest sites and prey availability may not coincide in urban areas for some species. Swainson's hawks (*Buteo swainsoni*) nesting in urban areas in California fledged fewer young than hawks in rural areas, possibly because the costs associated with foraging outside of the urban areas reduced nesting success (England et al. 1995).

Benefits to People

People are keenly interested in the wild animals that live around them, including in urban areas (see VanDruff et al. 1994 for a review), and frequently participate in activities designed to observe, feed, or photograph them (U.S. Fish and Wildlife Service and Bureau of the Census 1997). Birds of prey are, in the opinion of many people, charismatic and exciting to watch. For example, 80% of the residents of Tucson on or near whose property Cooper's hawks nested liked having the hawks nearby (Mannan et al., 2003). The primary reason given for liking the hawks was that they were interesting to watch. Thus, it is not surprising that predatory birds in urban areas have aesthetic value and often are looked upon with favor.

The presence of birds of prey in urban landscapes also has great potential as a means to educate people about the behavior and life history of predatory species, and about predator-prey interactions. Perhaps the best illustration of the educational value of predatory birds is the pair of peregrine falcons that nested on the United States Fidelity and Guaranty Building in Baltimore in 1977 (Cade and Bird 1990). Documentation and publicity of their activities introduced many people in that city (and beyond) to the nesting cycle and behavior of this species (Cade and Bird 1990). But educational opportunities associated with birds of prey also can take place on a small scale. For example, when we band nestling Cooper's hawks in Tucson, it is not uncommon for multiple families from the neighborhood around the nest to gather and watch the process. These gatherings present a wonderful opportunity for us to answer questions about the behavior of Cooper's hawks and their relationship with urban landscapes. Similarly, when nests of predatory birds are close to neighborhood schools, opportunities are abundant to involve classrooms with projects about birds of prey. Activities of this kind can foster a sense of "protective ownership" among residents who live close to the birds of prey.

Problems for Birds of Prey

Modification of natural environments by development often makes them unsuitable for occupancy by birds of prey. Sometimes, however, the habitat features that attract predatory birds remain intact, or are even enhanced, after development. The consequence is that some species attempt to reside in places that are different from those in which they evolved. Such places can present "environmental challenges" to which birds are not well adapted, and result in high rates of mortality. Electrocution, poisoning, disease, and collisions with windows or vehicles are examples of agents of mortality that can affect birds of prey in urban areas.

Electrocution of birds of prey on utility poles is a significant management problem in a variety of landscapes (Lehman 2001). Electrocutions can occur, for example, when an energized wire on a utility pole is positioned so that a bird can touch it and a ground wire simultaneously. Large birds, therefore, including many species of predatory birds, are more susceptible to electrocution than small birds. The high number of configurations that could electrocute birds in urban settings (e.g., transformers) makes these environments especially dangerous when predatory birds use them (e.g., Dawson and Mannan 1994).

Poisoning of birds of prey in urban landscapes has not been widely reported, but it is a potentially serious problem. For example, predatory birds could be exposed to toxic substances (e.g., strychnine, lead, and organophosphates) when they eat animals targeted by poisons or shooting (Cade and Bird 1990). Several peregrine falcons in urban areas have died from eating strychnine-poisoned pigeons (i.e., rock pigeons [*Columba livia*]; Cade and Bird 1990). Similarly, some Cooper's hawks in Tucson may die each year from organophosphate poisoning when they eat poisoned pest species (Boal and Mannan 1999).

Disease also may present a problem for predatory birds in some urban areas. In Tucson, the protozoan disease trichomoniasis kills about 40% of nestling Cooper's hawks annually (Boal et al. 1998; Boal and Mannan 1999). The nestlings probably contract the disease via their diet. Mourning doves (*Zenaida macroura*), white-winged doves (*Z. asiatica*), and Inca doves (*Columbina inca*) are among the most abundant species in Tucson (Germaine et al. 1998). These species carry the protozoan (*Trichomonas gallinae*) that causes the disease (Hedlund 1998) and comprise 57% of the prey items delivered by male Cooper's hawks to their nestlings (Estes and Mannan 2003).

Estimates of the number of birds in the United States killed annually by colliding with plate-glass windows range from 97.6 million to 975.6 million (Klem 1990). Birds of prey also are susceptible to this agent of mortality, and often collide with windows while chasing other birds. About 70% of the annual mortality of immature and adult Cooper's hawks in Tucson is caused by collisions with windows or moving vehicles (Boal and Mannan 1999). Similarly, collision with windows was identified as a serious hazard for urban-nesting peregrine falcons (Cade and Bird 1990).

The mortality agents associated with urban areas could be called "anthropogenic ecological traps" (e.g., Gates and Gysel 1978), because birds are triggered to settle in places that are dangerous to them. If the agent of mortality is widespread and severe, urban areas also could be "population sinks" (Pulliam 1988), or areas where reproduction does not balance mortality. Persistence of a population in a population sink obviously requires immigration from environments outside of the area. Whether mortality associated with urban areas is severe enough to create population sinks for any birds of prey is unclear and requires considerable information about population dynamics. A model of the dynamics of the population of Cooper's hawks in Tucson suggested that it was declining at about 8% per year, primarily because of trichomoniasis (Boal 1997). However, the primary weakness of the model was the estimate of survival of hawks between fledging and 1 year of age (Boal 1997). Subsequent estimates of the survival of juvenile Cooper's hawks in Tucson (Mannan et al., 2004) are higher than what was used in the model, and may alter conclusions about whether Tucson is a "source" or "sink." Nevertheless, the influence of "environmental challenges" in some urban areas on populations of predatory birds can be dramatic. For example, breeding groups of Harris's hawks (*Parabuteo unicinctus*) in Tucson are smaller in size and less stable than those occupying undeveloped desert, largely owing to the electrocution of adults and fledglings (Dawson and Mannan 1994).

Another problem for birds of prey in urban landscapes is intentional harassment by people. Despite laws against discharging firearms within developed areas, birds of prey are occasionally shot. Some probably are killed because they are viewed as nuisances. For example, 28 Mississippi kites were shot in 1978 in Ashland, Kansas, because they were diving at people (Parker 1996). Other predatory birds are killed because they are in the wrong place at the wrong time and happen to be large enough for a young boy with a pellet gun to hit. Harassment also can result from the actions of people who have good intentions but are uninformed. For example, young hawks and owls occasionally spend one to several days on or near the ground immediately after they fledge. Residents who see newly fledged birds on the ground frequently "rescue" them from real or perceived dangers and take them to wildlife rehabilitators. If the rehabilitators hold the young birds for several days or weeks, their chances of survival, even if replaced in the nest area, may be diminished if their flight muscles are not well developed or if the parent hawks have left the area.

Problems for People

One problem for people who live in close association with birds of prey is that individual birds of some species become aggressive when defending their nests and young, often diving over and even striking people when they walk under a nest or close to a fledgling (see review by Parker 1999). The best-documented cases of this behavior are with Mississippi kites, although it has been reported in a variety of species (e.g., great horned owl [*Bubo virginianus*], Cooper's hawks; see Parker 1999). Other concerns people have about birds of prey living near them can be categorized as nuisance factors, and include the mess made by the accumulation of fecal material and prey remains, predatory acts (real or perceived) on pets or favorite native animals, and noise associated with vocalizations for courtship or food begging early in the morning (Mannan et al., 2003).

A potentially significant economic problem for people can arise if a species, listed as threatened or endangered under the Endangered Species Act (ESA), is present in areas undergoing development. This problem is not limited to birds, but predatory birds often are the focus of management concerns, and their presence in urban areas has restricted development in some situations. For example, recommendations in the recovery plan for the cactus ferruginous pygmy-owl (*Glaucidium brasilianum cactorum*) limit development in areas occupied by the owl (U.S. Fish and Wildlife Service 2003), and the limitations can be costly from an economic perspective.

Management of Problems

Changes in management can help to ameliorate some of the problems described above, but others may be more intractable. Management issues can be divided into two basic groups, those that address the bird perspective and those that address the human perspective.

Managing the Problems for Birds of Prey

Some factors associated with urban landscapes that cause problems for birds of prey, such as configurations on utility poles that electrocute large birds, can be corrected. Considerable effort has been made to identify ways to modify or "retrofit" dangerous configurations to make them "raptor safe" (APLIC 1996). Materials required for modifications usually are relatively inexpensive, but installation is time- and labor-intensive. Moreover, configurations that cause electrocutions can be numerous and widespread; thus, in some situations, considerable time and resources will be required to remedy the problem. Proactively modifying dangerous poles in areas where birds of prey are active (e.g., near nests or in areas where they forage) may be the best short-term strategy in situations of this kind (Harness 2003).

The number of birds killed by window strikes could be reduced in several ways: (1) by moving objects that attract songbirds (and the birds that eat them), such as bird feeders, birdbaths, and vegetation providing cover or food, well away from windows; (2) by covering problem windows completely with netting or other material when cost and aesthetics permit; and (3) by angling windows downward on newly constructed buildings so that they reflect the ground instead of surrounding vegetation and sky (Klem 1990). Noteworthy is that single objects, such as falcon silhouettes, do not reduce window strikes because they do not cover enough of the window (Klem 1990).

Some factors associated with urban landscapes that cause problems for birds of prey may not be easy to modify; the incidence of disease may be one of these. For example, there probably is little that can or should be done to reduce the abundance of native dove species in Tucson, the rates at which they carry disease agents, the proportion of doves in the diet of Cooper's hawks, or the rates at which nestling Cooper's hawks die from the disease. Thus, this urban-related problem likely will persist unless it abates through natural processes. Fortunately, problems with disease may not be universal among urban areas. The incidence of trichomoniasis in nestling Cooper's hawks is relatively low in urban and rural areas of Wisconsin, North Dakota, and British Columbia (Rosenfield et al. 2002).

Problems for birds of prey associated with unintentional harassment or poisoning by people can be solved, at least partly, by education. For example, most people will leave fledgling hawks and owls they find on the ground alone if informed that the parent birds will care for them. (It should be noted, however, that when peregrine falcon nestling/fledglings leave the nest and flutter to the ground in urban areas, they are at great risk, and may not be cared for by their parents [Cade and Bird 1990]). Careful monitoring and control of the use of poisons that could kill nontarget species, combined with education about the potential effects of these toxins, may reduce the number of poisoned raptors. In contrast, solving problems associated with intentional harassment likely will be difficult because the people who harm birds generally know or suspect that their acts are illegal, or feel that their actions are justified (see below).

Managing the Problems for People

People usually are frightened or angry when predatory birds dive at their heads and strike them, or dive at and strike their children or pets. But their fear and anger sometimes are reduced once they understand that the birds are simply trying to protect their offspring and that serious injury as a result of these attacks is rare. Some people are even willing to modify their own activities, for the relatively short period when birds are prone to this behavior, to reduce the "attacks." In some situations, however, the aggressive behavior of predatory birds is unacceptable, such as when it occurs immediately outside a place of business, or when multiple birds are involved. Transplanting nestlings to rural nests may be necessary in such situations (Parker 1999). Alternatively, modifying the structure of nest sites can sometimes trigger predatory birds to relocate nests in subsequent years. Structures supporting nests (e.g., trees) could be removed completely in the nonbreeding season, but extreme action of this kind may not always be necessary. For example, trimming trees in winter to make them less full usually triggers Cooper's hawks in Tucson to rebuild nests in different groves the following spring (Mannan, pers. observ.).

Education also may reduce the concerns of people about other problems caused by predatory birds (e.g., their messiness and noise), but not all of them. Changing a person's views about predation, for example, may be difficult because some people find such acts distasteful, despite understanding that the acts are necessary for the predator (Mannan et al., 2003). Similarly, if the presence of a predatory bird that is listed as endangered under the ESA restricts a person from developing their land, that person's views about the bird will likely be negative, regardless of level of biological understanding.

Conclusion

The presence of birds of prey in urban landscapes results in benefits and problems for both birds and people. The benefits of having predatory birds live close to people are substantial, and many of the problems can be overcome with time and resources, or through education. However, it is incumbent on biologists and city planners to understand and be prepared to deal with the problems, if the conditions that promote birds of prey to occupy urban landscapes are to be maintained or created. Motivation for spending time and resources on solving problems created by having predatory birds and people live close to each other could come from legal mandates (birds of prey are protected by law), or from a desire by city or state agencies to appease the concerns of individual citizens, but ideally motivation should flow from an ethic that promotes treating animals inhabiting urban landscapes as members of the community (Leopold 1949).

Literature Cited

Adams, L. W. 1994. *Urban Wildlife Habitats.* University of Minnesota Press, Minneapolis.

Anderson, D. E., and D. L. Plumpton. 2000. "Urban landscapes and raptors: A review of factors affecting population ecology," in *Raptors at Risk,* ed. R. D. Chancellor and B. U. Meyburg, 434–445. WGBP and Hancock House, Blaine, WA.

Avian Power Line Interaction Committee (APLIC). 1996. *Suggested Practices for Raptor Protection on Power Lines: The State of the Art in 1996.* Edison Electric Institute and Raptor Research Foundation, Washington, DC.

Beissinger, S. R., and D. R. Osborne. 1982. "Effects of urbanization on avian community organization," *Condor* 84:75–83.

Blair, R. B. 1996. "Land use and avian species diversity along an urban gradient," *Ecological Applications* 6:506–519.

Bloom, P. H., M. D. McCrary, and M. J. Gibson. 1993. "Red-shouldered hawk home-range and habitat use in southern California," *Journal of Wildlife Management* 57:258–265.

Boal, C. W. 1997. "An Urban Environment as an Ecological Trap for Cooper's Hawks." Ph.D. dissertation, University of Arizona, Tucson.

Boal, C. W., and R. W. Mannan. 1998. "Nest-site selection by Cooper's hawks in an urban environment," *Journal of Wildlife Management* 62:864–871.

———. 1999. "Comparative breeding ecology of Cooper's hawks in urban and exurban areas of southeastern Arizona," *Journal of Wildlife Management* 63:77–84.

Boal, C. W., R. W. Mannan, and K. S. Hudelson. 1998. "Trichomoniasis in Cooper's hawks from Arizona," *Journal of Wildlife Diseases* 34:590–593.

Cade, T. J., and D. M. Bird. 1990. "Peregrine falcons, *Falco peregrinus*, nesting in an urban environment: A review," *Canadian Field-Naturalist* 104:209–218.

Dawson, J. D., and R. W. Mannan. 1994. *Population Dynamics of Harris' Hawks in Urban Environments.* Report to Arizona Game and Fish Department, Heritage Program, Phoenix.

Emlen, J. T. 1974. "An urban bird community in Tucson, Arizona: Derivation, structure, and regulation," *Condor* 76:184–197.

England, A. S., J. A. Estep, and W. R. Holt. 1995. "Nest-site selection and reproductive performance of urban-nesting Swainson's hawks in the Central Valley of California," *Journal of Raptor Research* 29:179–186.

Estes, W. A., and R. W. Mannan. 2003. "Feeding behavior of Cooper's hawks at urban and rural nests in southeastern Arizona," *Condor* 105:107–116.

Gates, J. E., and L. W. Gysel. 1978. "Avian nest dispersion and fledging success in field-forest ecotones," *Ecology* 59:871–883.

Gehlbach, F. R. 1994. *The Eastern Screech Owl: Life History, Ecology, and Behavior in the Suburbs and Countryside.* Texas A&M University, College Station.

———. 1996. "Eastern screech owl in suburbia: A model of raptor urbanization," in *Raptors in Human Landscapes: Adaptations to Built and Cultivated Environments,* ed. D. M. Bird, D. E. Varland, and J. J. Negro, 69–74. Academic Press, San Diego.

Germaine, S. S., S. S. Rosenstock, R. E. Schweinsburg, and W. S. Richardson. 1998. "Relationships among breeding birds, habitat, and residential development in greater Tucson, Arizona," *Ecological Applications* 8:680–691.

Harness, R. E. 2003. *Avian Protection Plan—Lee County Electric Cooperative, Inc.* Lee County Electric Cooperative, Fort Myers, FL.

Hedlund, C. A. 1998. "*Trichomoniasis gallinae* in avian populations in urban Tucson." M.Sc. thesis, University of Arizona, Tucson.

Klem, D. Jr. 1990. "Collisions between birds and windows: Mortality and prevention," *Journal of Field Ornithology* 61:120–128.

Lehman, R. N. 2001. "Raptor electrocution on power lines: Current issues and outlook," *Wildlife Society Bulletin* 29:804–813.

Leopold, A. 1949. *A Sand County Almanac.* Oxford University Press, New York.

Mannan, R. W., and C. W. Boal. 2000. "Home range characteristics of male Cooper's hawks in an urban environment," *Wilson Bulletin* 112:21–27.

Mannan, R. W., C. W. Boal, W. J. Burroughs, J. W. Dawson, T. S. Estabrook, and W. S. Richardson. 2000. "Nest sites of five raptor species along an urban gradient," in *Raptors at Risk,* ed. R. D. Chancellor and B. U. Meyburg, 447–453. WGBP and Hancock House, Blaine, WA.

Mannan, R. W., W. A. Estes, and W. J. Matter. 2004. "Movements and survival of fledgling Cooper's hawks in an urban environment," *Journal of Raptor Research.* 38:26–34.

Mannan, R. W., W. W. Shaw, W. A. Estes, M. Alanen, and C. W. Boal. 2003. "A preliminary assessment of the attitudes of people towards Cooper's hawks nesting in an urban environment," in *Proceedings of the Fourth International Symposium on Urban Wildlife Conservation, Tucson, Arizona,* ed. W. W. Shaw, L. K. Harris, and L. VanDruff, 1–5 May 1999. College of Agriculture and Life Sciences, University of Arizona, Tucson.

Marzluff, J. M., F. R. Gehlbach, and D. A. Manuwal. 1998. "Urban environments: Influences on avifuana and challenges for the avian conservationist," in *Avian Conservation: Research and Management,* ed. J. M. Marzluff and R. Sallabanks, 283–299. Island Press, Washington, DC.

Morrison, M. L., B. G. Marcot, and R. W. Mannan. 1998. *Wildlife-Habitat Relationships,* 2nd edition. University of Wisconsin Press, Madison.

Parker, J. W. 1996. "Urban ecology of the Mississippi kite," in *Raptors in Human Landscapes: Adaptations to Built and Cultivated Environments,* ed. D. M. Bird, D. E. Varland, and J. J. Negro, 45–52. Academic Press, San Diego.

———. 1999. "Raptor attacks on people," *Journal of Raptor Research* 33:63–66.

Pulliam, H. R. 1988. "Sources, sinks, and population regulation," *American Naturalist* 132:652–661.

Rosenfield, R. N., J. Bielefeldt, J. L. Affeldt, and D. J. Bechmann. 1995. "Nesting density, nest area, reoccupancy, and monitoring implications for Cooper's hawks in Wisconsin," *Journal of Raptor Research* 29:1–4.

Rosenfield, R. N., J. Bielefeldt, L. J. Rosenfield, S. J. Taft, R. K. Murphy, and A. C. Stewart. 2002. "Prevalence of *Trichomoniasis gallinae* in nestling Cooper's hawks among three North American populations," *Wilson Bulletin* 114:145–147.

Smith, D. G., T. Bosakowski, and A. Devine. 1999. "Nest site selection by urban and rural great horned owls in the Northeast," *Journal of Field Ornithology* 70:535–542.

Sodhi, N. S., P. C. James, I. G. Warkentin, and L. W. Oliphant. 1992. "Breeding ecology of urban Merlins (*Falco columbarius*)," *Canadian Journal of Zoology* 70:1477–1483.

Stewart, A. C., R. W. Campbell, and S. Dickin. 1996. "Use of dawn vocalizations for detecting breeding Cooper's hawks in an urban environment," *Wildlife Society Bulletin* 24:291–293.

Temple, S. A. 1977. "Manipulating behavioral patterns of endangered birds: A potential management technique," in *Endangered Birds: Management Techniques for Preserving Threatened Species,* ed. S. A. Temple, 435–443. University of Wisconsin Press, Madison.

U.S. Fish and Wildlife Service and U.S. Bureau of the Census. 1997. *1996 National Survey of Fishing, Hunting, and Wildlife-Associated Recreation.* U.S. Department of the Interior and U.S. Department of Commerce, Washington, DC.

U.S. Fish and Wildlife Service. 2003. "Draft Recovery Plan for the Cactus Ferruginous Pygmy-owl (*Glaucidium brasilianum cactorum*)," *Federal Register* 68:1189.

VanDruff, L. W., E. G. Bolen, and G. L. San Julian. 1994. "Management of urban wildlife," in *Research and Management Techniques for Wildlife and Habitats*, ed. T. A. Bookhout, 507–530. The Wildlife Society, Washington, DC.

Wecker, S. C. 1963. "The role of early experience in habitat selection by the prairie deer mouse, *Peromyscus maniculatus bairdi*," *Ecological Monographs* 33: 307–325.

Wiens, J. A. 1985. "Habitat selection in variable environments: Shrub-steppe birds," in *Habitat Selection in Birds*, ed. M. L. Cody, 227–251. Academic Press, San Diego.

CHAPTER 6

Challenges in Conservation of the Endangered San Joaquin Kit Fox

Howard O. Clark Jr., Brian L. Cypher, Gregory D. Warrick, Patrick A. Kelly, Daniel F. Williams, and David E. Grubbs

The San Joaquin Valley of California is a region of immense biological diversity and includes a number of endemic species and unique biotic communities. California harbors more unique plants and animals than any other state, and its biotic communities also face intense pressures. With the human population of California increasing every year, there is always more and more demand on the remaining habitat. Undeveloped habitat the size of San Francisco is converted to residential or commercial use every six months (Motavalli 2002). The state of California lists 34 species of animals and 46 species of plants as having been extirpated since the 1880s. The combined state and federal lists of rare, threatened, or endangered plant and animal species in the state total 330, and there are more candidate species identified. In the 1780s, with a population numbering in the hundreds of thousands, California had an estimated 5 million acres of wetlands. After just two centuries of human population growth, the state has more than 32 million people and only about 454,000 acres of wetlands, a 90% loss (Motavalli 2002).

The San Joaquin Valley of central California exemplifies these trends of habitat loss in the face of human pressures. Because of profound habitat loss and degradation due mainly to agriculture and urbanization, numerous plant and animal species in the San Joaquin Valley are considered sensitive, and many are receiving formal federal and state protection. Separate conservation strategies for each of these species would be inefficient and socioeconomically impractical. Thus, use of the endangered San Joaquin kit fox (*Vulpes macrotis mutica*) as an "umbrella" species in conservation

6. *Challenges in Conservation of the Endangered San Joaquin Kit Fox*

and recovery efforts has been proposed (USFWS 1998). Because kit foxes require relatively large tracts of land to support viable populations, the assumption is that conservation efforts, particularly habitat conservation, on behalf of the kit fox, will benefit other species by facilitating recovery of listed species, precluding the need to list additional species, and conserving significant portions of several unique biotic communities (USFWS 1998).

In this chapter we discuss the conservation of the San Joaquin kit fox in the face of habitat alteration and another anthropogenic factor: interspecific competition from nonnative red foxes (*V. vulpes*). We also give some strategies for conserving San Joaquin kit foxes in human-dominated landscapes by utilizing red fox control and habitat protection.

Conservation of the San Joaquin Kit Fox

Kit foxes are relatively small members (1.7–3.0 kg) of the family Canidae (order Carnivora) that occur in the desert and desertlike habitats of the southwestern United States and northern Mexico. The San Joaquin kit fox is a genetically distinct subspecies (Mercure et al. 1993) that inhabits suitable habitat on the San Joaquin Valley floor and in the surrounding foothills of the coastal ranges, Sierra Nevada, and Tehachapi Mountains. Prior to 1930, kit foxes inhabited most of the San Joaquin Valley from southern Kern County north to eastern Contra Costa County and eastern Stanislaus County. By 1930 the kit fox range had been reduced by more than half, with the largest portion remaining in the western and southern portions of the valley. San Joaquin kit foxes currently persevere as three core and several satellite populations connected by migrating individuals. These populations are scattered across islands of natural land amid agriculture on the valley floor in Kern, Tulare, Kings, Fresno, Madera, San Benito, Merced, Stanislaus, San Joaquin, Alameda, and Contra Costa counties. They also occur in the interior basins and ranges in Monterey, San Benito, San Luis Obispo, and, possibly, Santa Clara counties; and in the upper Cuyama River watershed in northern Ventura and Santa Barbara counties and southeastern San Luis Obispo County. Figure 6.1 shows the current range of the San Joaquin kit fox.

This significant reduction in the abundance and range of San Joaquin kit foxes is primarily attributable to human-caused habitat loss and degradation, specifically agricultural, industrial, and urban development (USFWS 1998). Much of the species' remaining habitat is fragmented,

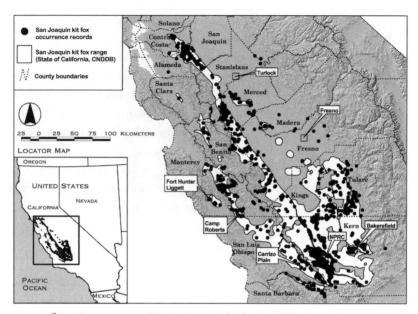

FIGURE 6.1 Current range of the San Joaquin kit fox in California. Fox locations originate from points maintained by the California Department of Fish and Game in the Natural Diversity Database. Created by Scott Phillips.

disturbed, and subject to competing land uses such as gas and oil production, water banking, and grazing (USFWS 1998). In addition, fur harvests, predator control programs, and rodent control programs also may have contributed to observed declines. For example, kit foxes and their close relative, the swift fox (*V. velox*), have been known to be the first carnivores at sites where poisoned carcasses were left out for wolves (*Canis lupus*) and coyotes (*C. latrans;* Bunker 1940; Kitchen et al. 1999), and both fox species are very susceptible to secondary poisoning from eating contaminated rodents (Schitoskey 1975; Littrell 1990). The San Joaquin kit fox was listed as federally endangered in 1967 and as California threatened in 1971. There are only a few thousand San Joaquin kit foxes, compared with the tens of thousands that probably existed prior to European colonization of California.

The recovery of San Joaquin kit foxes is guided by the *Recovery Plan for Upland Species of the San Joaquin Valley, California,* published by the U.S. Fish and Wildlife Service in 1998. The *Recovery Plan* uses sound scientific research in an attempt to delist most upland endangered species within a 20-year period in the San Joaquin Valley. The *Recovery Plan* calls for both ecosystem and community-level strategies for recovery of endangered species. In creating the community-level strategy, greater emphasis is placed on two groups of species because of their important roles in either conservation (umbrella species, like the kit fox) or ecosystem dynamics (keystone species, such as kangaroo rats). Umbrella species are those species whose protection will concomitantly protect other species that share the same habitats; "keystone species" are those whose presence is vital for the functioning of ecosystems and whose loss can impact a suite of other species.

With the implementation of the *Recovery Plan*, the San Joaquin kit fox can recover and be delisted from the Endangered Species list by conserving large blocks of land that can maintain robust, viable populations. However, there are many unknowns, such as how the kit fox interacts with other carnivores, both native and nonnative, and how the continued loss and fragmentation of habitat will affect the future survivability of the kit fox.

Interspecific Competition

Competition from other mammalian predators is an important factor affecting the remaining San Joaquin kit fox populations. Notably, coyotes and nonnative red foxes (*V. vulpes regalis*) compete with kit foxes for space, dens, and food. They also directly kill kit foxes as well as harass them (White et al. 1994, 1995). These competitive interactions have major implications for the conservation and recovery of San Joaquin kit foxes.

Red foxes constitute a potentially serious threat to San Joaquin kit foxes, especially because of their ability to readily adapt to modified habitats. Historically, after the Pleistocene epoch, native red foxes (*V. v. necator*) only occurred at high elevations in the Sierra Nevada and Cascade ranges in California, and therefore did not occur in any portion of the range of San Joaquin kit foxes (Grinnell et al. 1937; Kucera 1993). However, red foxes from the northern Great Plains, and possibly elsewhere, have been deliberately introduced by people into lower-elevation areas of California, notably the Sacramento Valley, since the 1890s. These foxes were brought in for hunting and trapping, or escaped from fur farms (Jurek 1992; Lewis et al. 1999). (According to Jurek [1992], *V. v. regalis* are a North American native fox endemic to boreal habitats of north-central United States and Canada; however, according to Kamler and Ballard [2002], *V. v. regalis* were introduced from Europe.)

Because of the incredible adaptability of red foxes, these nonnative individuals have spread rapidly and have colonized many regions of California, including the San Joaquin Valley. In addition, there may have been other subsequent red fox introductions in other parts of the San Joaquin Valley after 1900. Introduced red foxes have established breeding populations throughout the Sacramento Valley (Gray 1975, 1977). Gould (1980) reported the range expansion of this population into Contra Costa and Alameda counties, as well as additional sightings in Marin, Santa Cruz, Ventura, and Los Angeles counties.

Red foxes are able to survive in a multitude of environments, and occur in more habitats than any other canid (Samuel and Nelson 1990). Red foxes are able to survive in intermixed cropland, farmland, shrubland, mixed hardwood stands, and edges of open areas. They take particular advantage of edge habitats (Henry 1996a, 1996b), such as agricultural fields abutting against canals and aqueducts. Interactions with coyotes can drive red foxes to live closer to human habitation, another frontier that red foxes have adapted very well to (Dekker 1983; Baker and Harris 2000). Many human activities, such as agriculture, urban sprawl, and the building of canals and aqueducts, create suitable edges for red foxes to exploit (Henry 1996a, 1996b).

Red foxes are larger (3–8 kg) than kit foxes, and are very effective at competing with smaller canids. For instance, red foxes were used as biological control agents for arctic foxes on Alaskan islands, where red foxes possibly excluded arctic foxes from the best feeding areas, rather than outright killing them (Rudzinski et al. 1982; Bailey 1992). Because of their larger size, red foxes usually win competitive battles with kit foxes. Red foxes are known to have killed radio-collared kit foxes in at least two studies (Ralls and White 1995; Clark 2001). For example, during the month of December in 1998, a radio-collared kit fox was found dead in a drainage culvert along the California Aqueduct, in Kern County, California. The fox was discovered by two of the authors—it appeared that the fox was still alive after being attacked, and managed to crawl into the culvert to die. After the necropsy a few days later, it was determined that the fox was attacked by a red fox because the paired puncture wounds (associated with hemorrhaging) around the thoracic cavity measured 17–26 mm, indicative of a red fox kill (Disney and Spiegel 1992). Red foxes may also be displacing kit foxes in some locations by taking over habitat and dens (White et al. 2000). Along the California Aqueduct in Kern County, kit foxes primarily used ar-

6. Challenges in Conservation of the Endangered San Joaquin Kit Fox 123

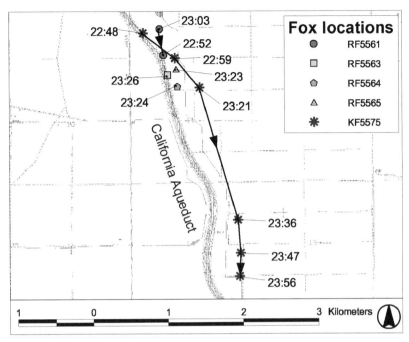

FIGURE 6.2 Avoidance travel by one kit fox to avoid four red foxes along the California Aqueduct in Kern County. The figure is temporally significant—the times show that the foxes were all in the same area around the same time. We cannot really know what the kit fox was thinking when it moved south from its earlier position. We can only speculate. The kit fox may have detected the reds in the area, and moved south—where its earthen den was located.

eas not occupied by red foxes, and kit foxes also vacated areas when red foxes approached (Clark 2001). During one instance during the aqueduct study (Clark 2001), a kit fox traveled 2 km in 30 minutes to avoid 4 red fox juveniles, as shown in Figure 6.2.

Although they are large enough to prevail over kit foxes in direct encounters, red foxes are close enough in size to exhibit significant niche overlap with kit foxes. This may intensify competition between these two species. Compared with other kit fox competitors, red foxes may have greater overlap in food item use because red foxes consume many of the same foods as kit foxes. Overlap in use of foods, shown in Figure 6.3, has been documented in the California Aqueduct location in Kern County (Clark 2001). Red foxes also will use kit fox dens, thus excluding use by kit foxes. Red foxes have been observed in dens formerly used by kit foxes on

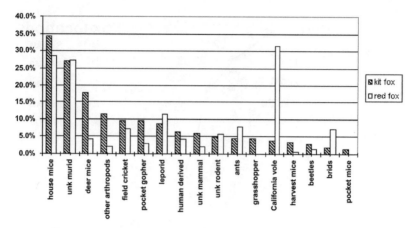

FIGURE 6.3 Food item use by kit foxes and red foxes at Lost Hills, California, in 1999. Bars are the proportions of scats (feces) with each food item. Scats were collected at known dens and other locations, and analyzed using hair and bone keys.

a number of occasions at two different locations (White et al. 2000; Clark 2001). Red foxes and kit foxes are also closely related taxonomically (Thornton et al. 1971; Creel and Thornton 1974), which may increase the potential for disease transmission (Cypher et al. 2001), although kit foxes may have a higher chance of contracting diseases from skunks and raccoons (White et al. 2000). However, red foxes represent the most widespread reservoir of rabies in Europe, Canada, and the United States (Lariviere and Pasitschniak-Arts 1996).

Adaptations of kit foxes that reduce competition with other competitors, such as using more than 30 escape dens year round, are less effective against red foxes. Kit foxes are dependent on dens both for microclimate control in the arid to semiarid regions they prefer, and for security from larger carnivores (Egoscue 1962, 1975). Kit foxes are always associated with these dens, unless they are hunting, which usually occurs during the nighttime hours. These dens may be a limiting factor in the distribution of kit foxes in California (Morrell 1972). Red foxes can enter most dens used by kit foxes, making escape and avoidance more difficult.

Some degree of habitat partitioning may occur between the two species when it comes to water. Kit foxes are able to use metabolic water from prey items (Golightly and Ohmart 1984) and do not need to drink directly from a water source, although they will if given the chance (H. Clark, pers. observ.). Red foxes, on the other hand, require a free source of water, and

6. Challenges in Conservation of the Endangered San Joaquin Kit Fox

most red fox observations in the San Joaquin Valley are in relatively close proximity to sources of water. This presents a limitation to red fox populations in some semiarid areas, such as the Carrizo Plain in San Luis Obispo County, California. However, the large number of man-made water sources in the San Joaquin Valley (e.g., canals, aqueducts, irrigated agriculture, stock ponds, and urban areas) has encouraged red fox colonization, exemplifying how other human activities can interact to promote the spread of a nonnative species at the expense of native wildlife.

Another canid species, the coyote, also interacts with kit foxes. The two species sometimes interact competitively, with the kit fox generally "losing" in the interaction. If given the chance, a coyote will run a kit fox down and kill it, usually by crushing the thoracic cavity and skull. The coyote usually does not eat the kit fox, but simply leaves the lifeless body of the fox where the attack occurred. It is not entirely known why coyotes kill kit foxes—the best hypothesis is that it is natural for larger canids to kill smaller canids to free up more resources for the domineering survivors. The same sort of relationship occurs between wolves and coyotes (Carbyn 1982). This phenomenon is not unique to North American carnivores; studies have shown that this fatal pecking order also occurs in Africa and elsewhere (Kitchen et al. 1999).

In general, however, kit foxes and coyotes can occupy the same habitats and share each other's territories. Kit foxes have co-evolved with coyotes and have developed mechanisms to enhance their survival, such as resource partitioning and greater dietary breadth, facilitating a level of coexistence (Cypher and Spencer 1998). Coyotes, for example, consume larger-sized food items, like rabbits, whereas kit fox consume smaller prey items, like nocturnal rodents (White et al.1995). Although coyotes and kit foxes have overlapping diets, both species exhibit specialization for prey items that reflect their specific energetics and body sizes, indicating a significant amount of resource partitioning and greater dietary breadth (White et al.1995).

Other research by White and colleagues (1994) determined that kit foxes did not avoid coyote-occupied areas. The home ranges of kit foxes significantly overlapped the home ranges of coyotes. Kit foxes did not refrain from using the overlapped portions of the interspecific ranges. The researchers observed coyotes and kit foxes foraging less than 200 meters from each other on three occasions. They hypothesized that avoidance behavior may only occur when the distance between the two species is even smaller than 200 meters. They also found that kit foxes use more than 30 dens throughout the year, providing a quick escape from approaching coyotes. Hence, kit foxes and coyotes can be sympatric (White et al. 1994).

More importantly for kit foxes, coyotes significantly impact red foxes via competition. Coyotes interact with and exclude red foxes at a much higher rate than they do with kit foxes. Reduced abundance of red foxes attributable to coyotes has been documented in a number of locations (Dekker 1983; Voigt and Earle 1983; Major and Sherburne 1987; Sargeant et al. 1987). This reduction is a consequence of both direct mortality and exclusion. For example, Sargeant et al. (1987) found that the abundance and distribution of the coyote affects the abundance and distribution of the red fox. The coyotes in this study (near the Missouri River, North Dakota) occupied larger home ranges than red foxes, and red foxes tended to avoid coyotes at all costs. The results were fewer red foxes in areas where coyotes are abundant. Voigt and Earle (1983) had similar findings in their study of coyotes and red foxes in Ontario, Canada.

During the California Aqueduct study, 11 radio-collared red foxes were recovered dead. The cause of death was identified for 8 of the 11 red foxes discovered; coyotes were responsible in all cases (Clark 2001). Therefore, although coyotes may compete with San Joaquin kit foxes, their presence actually may benefit kit foxes by reducing or suppressing red fox populations.

Strategies for Conserving San Joaquin Kit Foxes

Long-term conservation of the kit fox in the human-dominated landscape of the San Joaquin Valley will entail a two-pronged strategy: control of red foxes and securing the cooperation of landowners to protect the remaining kit fox habitat.

Red Fox Control

Implementing effective control programs for red foxes would be extremely difficult for a number of reasons. Poisons can not be used because of the threat to kit foxes as well as other nontarget species. Likewise, trapping red foxes without also capturing kit foxes would be difficult because of the relatively close weights of the two species. Also, the uses of many types of trapping devices were banned in California in 1998 by Proposition 4. Protect Pets and Wildlife (ProPAW) sponsored California's Proposition 4, formally known as the Wildlife Protection Act. The ProPAW initiative protects wildlife and family pets by banning indiscriminate traps, including the steel-jawed leghold trap and soft-catch legholds, for recreation or the fur trade. It also bans dangerous poisons that are harmful to animals and the environment, such as Compound 1080 (sodium fluoroacetate) and sodium cyanide, commonly administered through M-44s (spring-loaded devices that eject powdered sodium cyanide into the mouth of animals that bite down on scented bait. Upon contact with the animal's saliva, the powder generates cyanide gas that then kills the animal). Exceptions are provided to control nuisance animals and to protect public health and safety. However, no provisions were made to allow wildlife managers to use legholds to manage species that threaten endangered species.

Shooting, possibly in conjunction with predator calling (attracting a carnivore to the hunter or biologist by mimicking the sounds of prey species in distress), may be possible but could be difficult and costly to implement over large areas or near human-inhabited areas where red foxes commonly occur. Predator control programs are generally costly, as they usually must be implemented over large areas for multiple years (sometimes indefinitely) to achieve effective control. They are often unpopular with the general public even when conducted for the conservation of a rare, native species (Goodrich and Buskirk 1995). Thus, any reduction or limitation of red fox abundance achieved naturally through competitive pressure from native predators, such as coyotes, could significantly benefit kit foxes and would require no effort on the part of humans. Red foxes are rarely observed in areas where coyotes are abundant, even though kit foxes persist in these areas (Ralls and White 1995; Spiegel and Disney 1996; Cypher et al. 2000).

Coyotes have been suggested as a biological control strategy for red foxes in other regions. In coastal areas of California, the foxes are preying on endangered California least terns (*Sterna antillarum browni*) and California light-footed clapper rails (*Rallus longirostris levipes;* Jurek 1992) and are considered a threat to the salt marsh harvest mouse (*Reithrodontomys raviventris*) and Belding's savannah sparrow (*Passerculus sandwichensis beldingi*) (USFWS and U.S. Navy 1990). Coyotes also have been proposed as a means of reducing red foxes in the prairie pothole region of North America, thereby reducing red fox predation on duck nests (Sargeant and Arnold 1984).

However, there may be certain situations in which control of both coyotes and red foxes might benefit kit foxes. Such situations might include reintroduction sites, smaller preserves and habitat blocks during periods of low food availability, and specific locations where red foxes have become abundant. The last scenario may result from local coyote control efforts in the San Joaquin Valley that are conducted by federal and private personnel for livestock protection. In these situations, control would occur over a limited area and time period but could benefit kit foxes. None of these scenarios is being practiced at this time.

Habitat Protection

Federal safe harbor programs may be an important tool in the conservation of kit foxes, as they are for other species. In safe harbor agreements, landowners agree to undertake management actions to benefit an imperiled or endangered species on their property, and, in return, receive "safe harbor" assurances limiting Endangered Species Act responsibilities to those existing at the time the agreement is initiated ("the baseline" data determined by biologists). The safe harbor concept was developed by the environmental group Environmental Defense and the U.S. Fish and Wildlife Service to encourage private landowners to restore and maintain habitat for endangered species without fear of incurring additional regulatory restrictions (USFWS and NOAA1999).

In one of the first safe harbor agreements in California, the Paramount Farming Company agreed to install 25 artificial kit fox dens on its property near the California Aqueduct, in Kern County, for a minimum of three years. The objective is to allow kit foxes to move safely across the agricultural lands by using the dens as a method of escape from approaching coyotes and red foxes (USFWS 2001). Although no natural kit fox dens occur on Paramount's agricultural lands, kit foxes do den in grasslands to the immediate east and west of those lands. So far, biologists have found kit fox scats (genetically confirmed at the Smithsonian Institution's Molecular Genetics Laboratory, Washington, D.C.) near one of the dens, confirming that kit foxes are at least exploring the den complex. A young kit fox was trapped using a live box-trap in early 2003 near the series of dens. Unfortunately, the kit fox was killed by a coyote only 10 days after it was radio collared. Subsequent spotlighting efforts have shown that several other foxes live in the area, especially near the Bureau of Land Management parcels nearby. Monitoring of this project will continue.

Conclusion

Human development and agriculture have reduced and fragmented kit fox habitat and have also impacted the species by promoting increased abundance of introduced red foxes. Ultimately, the strategy with the greatest potential for effectively conserving and recovering San Joaquin kit foxes will be to conserve and properly manage large blocks of habitat that are connected by movement corridors, with few open water sources, and perhaps with artificial kit fox dens scattered throughout. These conservation efforts will facilitate larger kit fox populations that are more robust to losses from competition, and that are able to naturally repopulate areas where local extirpations of kit foxes may occur. The conservation of large blocks of habitat is a paramount goal of the recovery plan for San Joaquin kit foxes and a number of other rare species that occur in the same geographic range as kit foxes (U.S. Fish and Wildlife Service 1998).

Literature Cited

Bailey, E. P. 1992. "Red foxes, *Vulpes vulpes*, as biological control agents for introduced arctic foxes, *Alopex lagopus*, on Alaskan islands," *Canadian Field-Naturalist* 106:200–205.

Baker, P. J., and S. Harris. 2000. "Interaction rates between members of a group of red foxes (*Vulpes vulpes*)," *Mammal Review* 30:239–242.

Bunker, C. D. 1940. "The kit fox," *Science* 92:35–36.

Carbyn, L. N. 1982. "Coyote population fluctuations and spatial distribution in relation to wolf territories in Riding Mountain National Park, Manitoba," *Canadian Field-Naturalist* 96:176–182.

Clark, H. O. Jr. 2001. "Endangered San Joaquin kit fox and non-native red fox interspecific competitive interactions." M.S. thesis, California State University, Fresno.

Creel, G. C., and W. A. Thornton. 1974. "Comparative study of a *Vulpes fulva–Vulpes macrotis* hybrid fox karyotype," *Southwestern Naturalist* 18:465–468.

Cypher, B. L., H. O. Clark Jr., P.A. Kelly, C. Van Horn Job, G. D. Warrick, and D. F. Williams. 2001. "Interspecific interactions among mammalian predators: Implications for the conservation of endangered San Joaquin kit foxes," *Endangered Species Update* 18:171–174.

Cypher, B. L., and K. A. Spencer. 1998. "Competitive interactions between coyotes and San Joaquin kit foxes," *Journal of Mammalogy* 79:204–214.

Cypher, B. L., G. D. Warrick, M. R. M. Otten, T. P. O'Farrell, W. H. Berry, C. E. Harris, T. T. Kato, P. M. McCue, J. H. Scrivner, and B. W. Zoellick. 2000. "Population dynamics of San Joaquin kit foxes at the Naval Petroleum Reserves in California," *Wildlife Monographs* 145.

Dekker, D. 1983. "Denning and foraging habits of red foxes, *Vulpes vulpes*, and their interaction with coyotes, *Canis latrans*, in Central Alberta, 1972–1981," *Canadian Field-Naturalist* 97:303–306.

Disney, M., and L. K. Spiegel. 1992. "Sources and rates of San Joaquin kit fox mortality in western Kern County, California," *Transactions of the Western Section Wildlife Society* 28:73–82.

Egoscue, H. J. 1962. "Ecology and life history of the kit fox in Tooele County, Utah," *Ecology* 43:481–497.

———. 1975. "Population dynamics of the kit fox in western Utah," *Bulletin of the Southern California Academy of Science* 74:122–127.

Golightly, R. T. Jr., and R. D. Ohmart. 1984. "Water economy of two desert canids: Coyote and kit fox," *Journal of Mammalogy* 65:51–58.

Goodrich, J. M., and S. W. Buskirk. 1995. "Control of abundant native vertebrates for conservation of endangered species," *Conservation Biology* 9:1357–1364.

Gould, G. I. 1980. *Status of the Red Fox in California*. California Department of Fish and Game, Nongame Wildlife Investigation, Job I-8, progress report.

Gray, R. L. 1975. *Sacramento Valley Red Fox Survey*. California Department of Fish and Game, Nongame Wildlife Investigation. Progress report.

———. 1977. "Extensions of red fox distribution in California," *California Fish and Game* 63:58.

Grinnell, J., J. S. Dixon, and J. M. Linsdale. 1937. *Fur-Bearing Mammals of California*, vol. 2. University of California Press, Berkeley and Los Angeles.

Henry, J. D. 1996a. *Living on the Edge: Foxes*. NorthWord Press, Minocqua, WI.

———. 1996b. *Red Fox: The Catlike Canine*. Smithsonian Institution Press, Washington, DC.

Jurek, R. M. 1992. *Nonnative Red Foxes in California*, Nongame Bird and Mammal Section Report 92-04, State of California, Department of Fish and Game Wildlife Management Division.

Kamler, J. F., and W. B. Ballard. 2002. "A review of native and nonnative red foxes in North America," *Wildlife Society Bulletin* 30:370–379.

Kitchen, A. M., E. M. Gese, and E. R. Schauster. 1999. "Resource partitioning between coyotes and swift foxes: Spaces, time, and diet," *Canadian Journal of Zoology* 77:1645–1656.

Kucera, T. E. 1993. "The Sierra Nevada red fox," *Outdoor California* 54:4–5.

Lariviere, S., and M. Pasitschniak-Arts. 1996. "*Vulpes vulpes*," Mammalian Species 537:1–11. American Society of Mammalogists.

Lewis, J. C., K. L. Sallee, and R. T. Golightly Jr. 1999. "Introduction and range expansion of nonnative red foxes (*Vulpes vulpes*) in California," *American Midland Naturalist* 142:372–381.

Littrell, E. E. 1990. "Effects of field vertebrate pest control on nontarget wildlife (with emphasis on bird and rodent control)," in *Proceedings of the 14th Vertebrate Pest Conference*, ed. L. R. Davis and R. E. Marsh, 59–61. University of California, Davis.

Major, J. T., and J. A. Sherburne. 1987. "Interspecific relationships of coyotes, bobcats, and red foxes in western Maine," *Journal of Wildlife Management* 51:606–616.

Mercure, A., K. Ralls, K. P. Koepfli, and R. K. Wayne. 1993. "Genetic subdivisions among small canids: Mitochondrial DNA differentiation of swift, kit, and arctic foxes," *Evolution* 47:1313–1328.

Morrell, S. 1972. "Life history of the San Joaquin kit fox," *California Fish and Game* 58:162–174.

Motavalli, J. 2002. "As its population soars, California's environment approaches a crisis," *E/The Environmental Magazine*, March 21.

Ralls, K., and P. J. White. 1995. "Predation on San Joaquin kit foxes by larger canids," *Journal of Mammalogy* 76:723–729.

Rudzinski, D. R., H .B. Graves, A. B. Sargeant, and G. L. Storm. 1982. "Behavioral interactions of penned red and arctic foxes," *Journal of Wildlife Management* 46:877–884.

Samuel, D. E., and B. B Nelson. 1990. "Foxes," in *Wild Mammals of North America: Biology,*

6. Challenges in Conservation of the Endangered San Joaquin Kit Fox

Management, and Economics, 2nd edition, ed. G. A. Feldhamer, B.C. Thompson, and J. A. Chapman, 475–490. John Hopkins University Press, Baltimore.

Sargeant, A. B., S. H. Allen, and J. O. Hastings. 1987. "Spatial relations between sympatric coyotes and red foxes in North Dakota," *Journal of Wildlife Management* 51:285–293.

Sargeant, A. B., and P. M. Arnold. 1984. "Predator management for ducks on waterfowl production areas in the northern plains," in *Proceedings of the 11th Vertebrate Pest Control Conference,* 161–167. University of California, Davis.

Schitoskey, F. Jr. 1975. "Primary and secondary hazards of three rodenticides to kit fox," *Journal of Wildlife Management* 39:416–418.

Spiegel, L. K., and M. Disney. 1996. "Mortality sources and survival rates of San Joaquin kit fox in oil-developed and undeveloped lands of southwestern Kern County, California," in *Studies of the San Joaquin Kit Fox in Undeveloped and Oil-Developed Areas,* ed. L. K. Spiegel, 71–92. California Energy Commission, Sacramento.

Thornton, W. A., G. C. Creel, and R. E. Trimble. 1971. "Hybridization in the fox genus *Vulpes* in west Texas," *Southwestern Naturalist* 15:473–484.

U.S. Fish and Wildlife Service (USFWS). 1998. *Recovery Plan for Upland Species of the San Joaquin Valley, California.* U.S. Fish and Wildlife Service, Portland, OR.

U. S. Fish and Wildlife Service (USFWS). 2001. "Availability of an environmental action statement and receipt of an application from Paramount Farming Company for a permit to enhance the survival of the San Joaquin kit fox in Kern County, CA," *Federal Register* 66:50444–50445.

U.S. Fish and Wildlife Service (USFWS) and National Oceanic and Atmospheric Administration (NOAA). 1999. "Announcement of final safe harbor policy," *Federal Register* 64:32717–32725.

U.S. Fish and Wildlife Service (USFWS) and U.S. Navy. 1990. *Endangered Species Management and Protection Plan, Naval Weapons Station—Seal Beach and Seal Beach National Wildlife Refuge.* Final environmental impact statement. Portland, OR.

Voigt, D. R., and B. D. Earle. 1983. "Avoidance of coyotes by red fox families," *Journal of Wildlife Management* 47:852–857.

White, P. J., W. H. Berry, J. J. Eliason, and M. T. Hanson. 2000. "Catastrophic decrease in an isolated population of kit foxes," *Southwestern Naturalist* 45:204–211.

White, P. J., K. Ralls, and R. A. Garrott. 1994. "Coyote-kit fox interactions as revealed by telemetry," *Canadian Journal of Zoology* 72:1831–1836.

White, P. J., K. Ralls, and C. A. Vanderbilt White. 1995. "Overlap in habitat and food use between coyotes and San Joaquin kit foxes," *Southwestern Naturalist* 40:342–349.

CHAPTER 7

Carnivore Conservation and Highways: Understanding the Relationships, Problems, and Solutions

Bill Ruediger

Peter Singleton, a research scientist for the USDA Forest Service in Wenatchee, Washington, noticed the dead animals and the barrier effect that Interstate 90 (I-90) was having on wildlife populations in the Snoqualmie Pass area in the North Cascade Mountains in northwestern Washington. In response to this problem, he developed a research project to determine how the interstate could be modified to provide better habitat and population connectivity. The Washington Department of Transportation was receptive to his proposal, and Singleton and a small group of researchers began to study how wildlife interface with I-90.

Interstate 90 is a busy highway that generally has two or three traffic lanes in each direction, and traffic volume is so high that more lanes are needed. There are eight lanes on top of the pass, and still there is not enough capacity on I-90 for present and future traffic flow. The interstate virtually cuts the Wenatchee and Baker-Snoqualmie national forests in half and impedes movement of many wildlife and fish species. Some of the carnivores adjacent to I-90 include black bear (*Ursus americanus*), mountain lion (*Puma concolor*), lynx (*Lynx canadensis*), wolverine (*Gulo gulo*), coyote (*Canis latrans*), raccoon (*Procyon lotor*), and American marten (*Martes americana*). Grizzly bear (*Ursus arctos*) and wolves (*Canis lupus*) are slowly moving south from Canada and may repopulate the northern Cascades—if they can move across the obstacles that inhibit their movements.

Singleton's study involved searching for wildlife tracks at bridges and culverts that cross I-90. The crew also set up cameras to document species approaching or using bridges and culverts. Their work involved checking both ends of the bridges and culverts. Immediately it became apparent that for researchers to cross the surface of I-90 to check the opposite end of a culvert was far too dangerous. Even with our superior intelligence, crossing the highway was an invitation to serious injury or death. Just standing on the roadside was dangerous. So what was the solution? Singleton and his research team agreed to get into the vehicles and go to the next exit, cross safely under the highway, and drive to the other side of the culvert. No crossing of the pavement would be allowed.

The challenges that the busy interstate highway posed to Singleton and his crew are faced by wildlife every day across the country. The effects of highways and roads on carnivores and other wildlife are serious. There is no longer much question about the impact of highways on carnivores; the only question is, How should we address these issues?

Carnivore Sensitivity to Highways

Carnivores have biological traits that contribute to their vulnerability to highways. These include low population densities, low reproductive rates, large (huge for some species) home range sizes, and a high level of mobility. Species with low population densities and large area requirements, such as carnivores, are usually the first species to be affected by highways. The large home ranges of most of mid-sized and large carnivores require them to cross highways regularly, and most must cross or interface with several highways to fulfill their biological needs. Many carnivores have low reproductive rates and population densities. These factors often result in human-caused mortality that is additive rather than compensatory. Human-caused mortality is known to have contributed to extirpation of, or adverse population affects on, grizzly bear, black bear (Gibeau and Heuer 1996), wolves (Paquet and Callahan 1996), wolverine, lynx (Koehler and Brittel 1990; Ferreras et al. 1992), and fisher (*Martes pennanti*) (Krohn et al. 1994). Carnivores often exhibit ecological stress and extirpation before impacts on other wildlife are obvious.

Landscapes required to sustain mid- and large-sized carnivore populations are immense, especially when we consider expanding human populations. Figure 7.1 provides an estimate of the ecosystem size necessary to sustain carnivores in the northern Rocky Mountains (Paquet 1995). This area extends from west-central Wyoming to mid–British Columbia and Alberta. To utilize all the habitat in this area, carnivores are required to cross

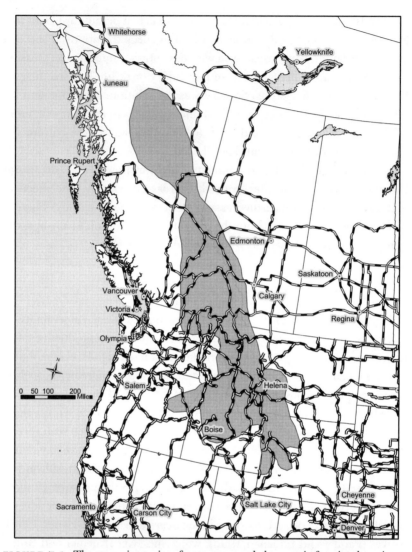

FIGURE 7.1 The approximate size of ecosystem needed to sustain functional carnivore populations in the northern Rocky Mountains. Although this area encompasses over 200,000 square miles, it is a fraction of any carnivore species' present or historic range. Modified from *Large Carnivore Conservation in the Rocky Mountains* (Paquet 1995).

at least 4 highways in Wyoming, 17 highways in Idaho (including 2 interstates), 23 highways in Montana (including 2 interstates), and 17 highways in British Columbia and Alberta (including the Trans-Canada Highway). Carnivores must contend with 61 highways in the northern Rocky Mountains (Ruediger 1996). This region is experiencing growth in human population, increased tourism, and rapidly accelerating commercial and resident traffic volumes. To meet these demands, highways are being upgraded and added to the system at an alarming rate. Similar impacts on carnivores and other wildlife are happening throughout the world.

How Highways Adversely Affect Carnivores

There are four ways in which highways negatively affect carnivores: direct mortality, direct habitat loss from road construction, habitat fragmentation, and the impacts of associated human development. The degree to which these factors affect carnivore conservation is just coming to light, but the impacts are severe—even where human population densities are relatively low.

Direct Mortality

Direct mortality from collisions with vehicles is the most obvious way that highways impact carnivores. Carnivore populations are particularly susceptible to highway mortality because of their large home ranges, long life spans, and low biological productivity. As mentioned previously, human-caused mortality has been a contributing factor in the decline of many carnivores.

Examples of direct highway mortality on wolves are numerous. In Banff National Park in 1996, 11 wolf mortalities were caused by collisions with cars, trucks, and trains. This number was equivalent to the entire known wolf reproduction for Bow River Valley (P. Paquet, pers. comm.). When wolves recolonized northwestern Montana, two alpha male wolves were killed on I-90. Vehicle collisions have become a regular occurrence for the wolf populations in Yellowstone National Park and central Idaho. In Weaver's wolf study in Jasper National Park (J.C. Weaver, pers. comm.). and Paquet's in Banff National Park, (P. Paquet, pers. comm.) highway and railroad mortality averaged 1–2 per pack per year.

Lynx highway mortalities have recently been observed in Colorado, where 10% of the known mortality of a reintroduced population has been vehicle collisions (Colorado Division of Wildlife 2000). Clevenger has also seen lynx carcasses on the Trans-Canada Highway. (T. Clevenger, pers. comm.). Sixteen of eighty-three translocated lynx were killed on highways in New York (Brocke et al. 1991). Ferreras et al. (1992) in their study of Iberian lynx (*Felis pardina*) in southwestern Spain found that the second most important cause of mortality was road traffic. Florida panthers (*Puma concolor coryi*) have been particularly vulnerable to direct mortality from road traffic. Harris and Gallagher (1989) have found that 65% of known Florida panther deaths since 1981 have been roadkills, while Maehr et al. (1991) calculated road mortality at a slightly more conservative figure of 49% of all documented panther deaths. In Texas, the major cause of mortality for the ocelot (*Felis pardalis*) population has been automobile collisions (Jenkins 1996).

Gibeau (1993) in his study of coyote use of urban habitats found 21 study animals killed on highways between July 1991 and March 1993. This constituted a 35% highway mortality rate. Gibeau and Heuer (1996) have compiled data for roadkills for various carnivores on the Trans-Canada Highway in the Bow River Valley. In Gibeau's coyote study, coyote roadkill was so high that it caused a concern that the study objectives could not be ascertained.

Information from Florida on black bear roadkills suggests that while vehicle use of highways in the study area increased by 100% (to 24,000 vehicle trips per day), black bear roadkills went from <5 to >90 from 1976 to 1995. (Gilbert and Wooding 1996). Increases in collisions with wildlife often occur at much higher rates than the increase in traffic volume. The result is that even small increases in traffic volume can have significant adverse affects on wildlife mortality.

Grizzly bears are rarely killed on highways, although documented fatal collisions have occurred on Highways 93, 83, and 2 in Montana, and on the Trans-Canada Highway before fencing was employed. The rare occurrences of grizzly mortality are likely due to their general avoidance of highways and their low population densities.

Wolverine highway mortality, as with grizzly bears, is rarely detected—probably for the same reasons. Wolverine are rare near major highways and probably avoid them. Dave Lewis (1996) observed wolverine mortality on highways and railroads in British Columbia (1 of 13 radio-collared wolverine was killed on a highway, and 1 on a railroad). John Krebs (pers. comm.) also has observed wolverine roadkills in Canada.

Direct Habitat Loss

Direct habitat loss is an obvious impact that is rarely documented. However, the cumulative effects of habitat loss must be staggering across North America and other continents. A 300-foot cleared right-of-way would consume 5.7% of each section it crosses. Indirect habitat loss due to displacement or wildlife avoidance of highways is unclear (Paquet and Callaghan 1996) but may be 1 km on each side of a highway in heavily forested or vegetated areas, and even greater distances in open habitats. This habitat loss should be considered permanent.

Millions of acres of habitat loss have occurred because of the direct and indirect effects of highways. Approximately 71 million acres of wildlife habitat have been lost under highways in the United States alone. Law requires mitigation of wetland habitat, and other infrastructures such as dams and power lines usually provide mitigation in the form of acquisition of habitat of equal quality and quantity to that affected by the projects. Unfortunately, mitigation of terrestrial habitat for highways is rarely done.

Habitat Fragmentation

Wilcox and Murphy (1985) have stated that "habitat fragmentation is the most serious threat to biological diversity and is the primary cause of the present extinction crisis." It is thus the most important issue to address and correct. Habitat fragmentation can be difficult to identify at local scales but can be easily seen on a map showing primary and secondary U.S. highways (Figure 7.2).

Carnivore habitat in southern Canada and the United States usually consists of islands of suitable habitat separated by marginal habitat or nonhabitat. The islands of suitable habitat are usually created by various vegetation patterns associated with geological features. In the west, these islands of habitat are often mountain ranges, or the valleys of major rivers. Most mountain ranges must be connected to many other mountain ranges to sustain an area large enough to maintain carnivore populations. However, valleys that separate mountain ranges are often occupied by humans, and most of the valley bottoms have highways and other human developments that make wildlife movement difficult or impossible (Weaver et al. 1996; Caroll et al. 2000). Highways and other human developments tend to create boundaries for both individuals and populations (see Harris and Gallagher 1989; Matthiae and Stearns 1981; Harris et al. 1982; Boecklen and Gotelli 1984; Maehr 1984; Noss 1991; Ferreras et al. 1992; Reed et al. 1996). These barriers fragment carnivore populations and ecosystems, which can result in the extirpation of small metapopulations. To ensure persistence over long periods of time, carnivore populations must optimally be large and be connected to other populations. The degree of population and habitat connectivity is inversely related to the number of highways

FIGURE 7.2 Map of the United States highway system, showing the maze of primary roads carnivores and other wildlife must deal with. Each highway represents a serious challenge, or barrier, to wildlife movement and wildlife habitat connectivity. This map provides a perspective of the main highway system only. Most rural roads and forest roads are not shown.

and other disturbance corridors that divide the landscape into disjointed pieces (Jalkotzy et al. 1997). Rare carnivores are generally present only in locations with the lowest highway densities.

Fragmentation of habitat begins when roadless areas are developed and progresses with the paving of gravel roads. As human presence increases, rural roads evolve into busy highways. Eventually highways can become barriers through a combination of increasing traffic volume and structural features such as concrete barriers, fences, and cut banks. Well before these physical features of roads form impenetrable barriers to wildlife, however, they influence carnivore behavior. Recent information from Yellowstone National Park (Mattson et al. 1987) and Banff National Park (Paquet et al.1994) suggests that wolves and grizzly bears are displaced by highways and generally avoid crossing them. Wildlife's behavioral avoidance of highways contributes in several ways to the fragmentation effect of highways.

CONSTRICTION OF HOME RANGES

Highways can fragment habitat by artificially constricting home ranges, often severing the most essential low-elevation feeding or breeding areas that would benefit an individual or population. An example is the effect of the Trans-Canada Highway in the Bow River Valley, where important low-elevation bottomland habitat is severed (perhaps only partially because of wildlife crossings). This fragmentation affects both the elk carrying capacity (prey) and the ability of wolves to effectively exploit their primary habitat (low-elevation gentle terrain). Additionally, in combination with the towns of Banff and Canmore, the highway restricts home ranges of wolf packs primarily to the upper Bow River Valley, while most of the prey species migrate east to the Rocky Mountain foothills in winter. This has led to a number of problems, including reduced elk densities in the upper Bow River Valley, the reduction of pack viability in Banff National Park, and unnaturally high densities of elk and other prey in the town site of Banff and the foothills (Paquet and Callaghan 1996). These researchers concluded that "wolves have been physically displaced, partially alienated, or blocked from using a minimum of 92 km^2 of the Bow River Valley's montane, i.e., 62% of the best wolf habitat in the Bow River Valley. Much of the problem is the result of disruption from the Trans-Canada Highway." Grizzly bear, black bear, and mountain lion are also severely impacted by the Trans-Canada Highway. Copeland (1994) and others (Gibeau and Heuer 1996) have noticed that wolverine and other carnivores' home ranges tend to be along highways, rather than extending across them.

DISRUPTION OF SEASONAL MOVEMENTS

Several carnivore species—lynx, wolverine, and wolves—are known to occasionally move long distances, sometimes well over 100 miles. Long-distance movements probably are part of the life strategies of many other wildlife species as well. In many situations, individuals leave their home ranges to find food or mates, then return months or weeks later. These temporary movements are not to be confused with dispersal out of the home range, which is usually permanent. Researchers found that lynx would leave their normal home range to exploit ground squirrels or other prey, and documented seasonal lynx movements of 20–30 km (Ruggiero et al. 2000). Squires also documented a male lynx in south-central Wyoming that moved several hundred miles to southern Montana, then returned. Seasonal long-distance movements of wolves (Mech and Boitani 2003) and wolverine (Copeland 1996) have also been observed. This phenomenon is probably common behavior for many carnivores and is impeded by highways.

BARRIERS TO DISPERSAL

Dispersal of young is critical need for population fitness and viability. Young animals, particularly males, usually must establish home ranges outside that of the parents. Long-range dispersal of young animals is common among large and mid-sized carnivores. Highways can act as filters or barriers to such movements, with the effect of loss of small metapopulations and reduction of overall population fitness (Singleton et al. 2002). For several forest carnivore species in the northern Rockies and Cascades (e.g., lynx, wolverine, and fisher), most of the populations consist of small groups of animals separated by many miles of secondary or marginal habitat, or non-habitat (Ruggerio et a. 1994). Dispersal for these animals can involve movements over long distances to find unoccupied or underoccupied suitable habitat. When highways block these movements, populations are fragmented into small populations that are more vulnerable to extinction.

Female home ranges are usually smaller than those of males in carnivore species, and female carnivores such as wolverine and bears often disperse to new home ranges over much smaller distances than males. Since population growth and range expansion is largely dependent on female dispersal and reproductive success, the ability for females to move uninhibited across potential habitat is an important conservation factor, particularly for species that are threatened or endangered. Highways are thought to be greater barriers for female grizzly bear (Gibeau and Heuer 1996) and wolverine (J. Krebs and J. Copeland, pers. comm.) than for males. In situations in which population expansion is necessary to meet recovery or conservation objectives, the inability of females to move across highways and other human barriers can result in failure to recover species. To compensate for this problem, agencies are often faced with costly and controversial translocation programs, such as occurred for wolves in Yellowstone and central Idaho, lynx in Colorado, and grizzly bears in the Cabinet-Yaak and central Idaho recovery areas.

DIFFERENTIAL AVOIDANCE AND CARNIVORE RECOVERY

Overall, most carnivores are intimidated by highways and tend to avoid them when possible (Jalkotzy et al. 1997). However, the same highway can affect different species, and individuals within the same species, in different ways. Wolverine and grizzly bear exhibit a high level of avoidance of human activities and of open roads and highways. One of the serious highway coordination issues with wolverine and grizzly bear will be to determine how to encourage them to approach and cross highways. Female grizzly bears and wolverine seem even more averse to crossing highways than males. Human activity, such as vehicle noise, may affect felids like lynx and mountain lion less. Marten, fisher, and river otter (*Lutra Canadensis*) readily use existing drainage culverts and may utilize these facilities to cross highways. Wolves and grizzly bear are very selective about using even large underpasses and overpasses designed for wildlife movement across highways. Coyotes and black bear appear to adjust to underpasses relatively quickly.

The solutions to the complex issues of wildlife habitat and population connectivity will not be "one size fits all." An issue with carnivore conservation that has not been addressed is the filter effect highways can have on certain age classes and sexes. The passage of a small number of animals can sometimes be perceived as "success." Often these animals are of the same age group and sex (sexually mature males). It is important to provide habitat connectivity for all ages and sexes, particularly dispersing females and females with young.

Associated Human Development

As access increases, the amount of associated development increases also. Land values reflect ease of access. In the Yaak area of northwestern Montana, the paving of what is now State Highway 508 and the increased ease of access have resulted in subdivisions and increased seasonal and yearlong human use of a once remote valley. The Yaak River Valley is home to grizzly bear, black bear, wolves, mountain lion, lynx, wolverine, fisher, American marten, and other carnivores. Whether these animals can persist along with the increased human use and development remains to be seen. This impact is severe and permanent for carnivore communities. There are many other examples of areas once highly suitable for carnivores that have been adversely impacted after major highway improvements (Ruediger et al. 1999). The assumption should be that if we build a better highway, they will come (humans). And, they do!

What Should Be Done to Mitigate and Restore Habitat Connectivity for Carnivores?

While the impacts of highways on carnivore populations are serious, both science and technology are helping to compensate for many adverse affects. The challenge ahead is to ensure that the available science and technology are applied on appropriate highways, and to continue progress through additional research and technology development. Without these measures, many areas that presently have large and mid-sized carnivores will likely see declines and possible extirpations. Wildlife habitat linkage analysis, building wildlife crossings, fencing highway right-of-ways, acquisition of key wildlife linkage areas, and developing and implementing clear agency policies are absolutely critical for carnivore populations to continue into the twenty-first century.

Wildlife Habitat Linkage Analysis

Linkage analysis is simply the mapping of large "core areas" where carnivores presently exist, or could exist, and ensuring that these habitats are connected over time. In many instances, serious habitat fragmentation has already occurred, and restoration of connectivity is needed. In much of Europe, habitat fragmentation occurred decades or centuries ago, and there are large international programs to restore habitat connectivity (Bank et al. 2002).

Wildlife habitat linkage analysis must be a cooperative effort between highway agencies, wildlife agencies, land management agencies, and conservation groups. There are a number of processes that can be used to make these assessments, such as those done in the Cascade Mountains by Singleton et al. (2002). In the northern Rocky Mountains, wildlife linkage analysis has been done by Walker and Craighead (1997); Ruediger, et al. (1999); Servheen, Waller, and Sandstrom (2001); and the Interagency Grizzly Bear Committee (Ruediger 2001).

Several other western states, including Montana, Idaho, Wyoming, Washington, and Colorado, have recently assessed habitat connectivity for lynx. New Mexico had a habitat connectivity meeting in June 2003 and has a broad-scale plan for maintaining wildlife habitat connectivity across the state. California has a conservation group–proposed habitat connectivity plan and is making progress toward implementation. More agency support is needed to make this a reality. Other western states also have interest in developing statewide habitat connectivity plans. Within the next few years, there will almost certainly be in place habitat connectivity plans for the Rocky Mountain region of the United States that provide continuous wildlife linkage from the Mexican border to the interior Canadian Rockies. Also, it is likely the Oregon Cascade Mountains will have connectivity with the mountains in northern California and Sierras. The Cascade Mountains in Washington could also have linkages with western Canada and the northern Rocky Mountains. Implementation of these plans will take millions or billions of dollars and serious commitments from all involved agencies.

The outlook for developing wildlife linkages in the Midwest heartlands and eastern states is not nearly so promising. The exception is Florida, which has an excellent plan and program for habitat linkage of Florida panther and black bears, as well as many other species (Cox and Kautz 2000). This is a critical plan for Florida because of the onslaught of humanity moving there every day. What is needed are similar plans for other Midwest and eastern states. These plans should provide wildlife habitat and population linkage within the states and between adjacent states. A great deal of thought and habitat protection and restoration is needed to ensure that eastern populations of black bear, cougar, wolves, American marten, river otter, bobcat, lynx, and fisher persist over the next century or two. Without a comprehensive program of wildlife habitat linkage analysis in the Midwest and eastern states, many or most of the present carnivore populations face imminent or eventual extirpation as the combination of habitat loss, habitat fragmentation, and increased human-caused mortality will apply their unrelenting forces.

Wildlife habitat linkage analysis is critical to maintaining intact and viable populations of carnivores. It is also essential to determine where wildlife crossings should be placed, how many are needed, and what type of structures are needed. Building wildlife crossings without forethought as to which habitats and populations need connectivity is an inefficient way to expend limited wildlife mitigation dollars. Wildlife habitat linkage analysis can also provide important information for reducing vehicle accidents with deer, elk, and other large animals; determining where open space is needed, and for assessing potential situations in which landownership adjustments and conservation easements are recommended to maintain overall ecosystem sustainability.

Building Wildlife Crossings

Once wildlife habitat linkage analysis has been completed, transportation agencies can begin to plan, design, and implement wildlife crossings. While the challenges of designing effective wildlife crossings for some species can be difficult, for most carnivores and other wildlife there are effective crossing designs with proven results (Clevenger 2000; Forman et al. 2003). The basic types of wildlife crossings include land bridges, multispans, open spans, steel culvert underpasses, bridge extensions, and various types of box and round culverts. Some wildlife crossing structures, such as land bridges and multispans, are effective for all species but are expensive to build. Other structures, such as large box culverts, are much more economical but may not be accepted by all species or sex and age classes. The best example of the latter problem is on the Trans-Canada Highway wildlife crossings in Banff. Grizzly bears, considered one of the most difficult species to make effective wildlife crossings for, prefer to cross the Trans-Canada Highway through either open spans or land bridges and generally shun other types of wildlife structures. Wolves prefer the same type of structures as grizzly bear but use steel culvert underpasses more often. Black bear, cougar, and coyote show preferences for 4 m round culverts, 3 m box culverts, or the 7×4 m steel culvert underpasses that grizzly bear and wolves either avoid or use sparingly (Clevenger and Waltho 1999; Clevenger 2000).

Some species have evolved using tree cavities or burrows and readily accept relatively small cement culverts. These include badger, (*Taxidea taxus*) mink and weasels (*Mustela* spp.), American marten, and fisher. Each situation needs to be analyzed according to the species present and the habitats they prefer to use. Putting the right kind of structures in the right place and having enough of them is the best ticket to success. As a general rule, a compromise between cost and acceptance by a wide variety of large carnivores seems to be the open-span wildlife crossing used in Banff National Park and elsewhere. Wildlife crossings benefit carnivores by reducing both habitat fragmentation and vehicle-caused mortality, both extremely important factors in conserving these species. Wildlife crossings also have the ancillary benefit of reducing vehicle collisions with all wildlife.

Fencing Highway Right-of-Ways

Even the best-designed wildlife crossings may fail without fencing of the highway right-of-way. While many wildlife crossing structures appear large and roomy by human standards, they are small and unnatural to many animals. Carnivores tend to be especially wary of anything unnatural and may cross the highway surface rather than use wildlife crossings. This, of course, negates the purpose of the crossing and can be fatal to the animals. To increase use of wildlife crossings, fencing is provided to guide or force animals into the crossing. The fencing can be either continuous, as on much of the Trans-Canada Highway, or wing fencing that extends for a half mile or so on each side of the crossings. Wing fencing may not prevent all animals from crossing the highway surface, but it may substantially reduce highway-caused mortality. If animals indicate a high level of avoidance to structures, either the fencing and/or the structures probably need modification. Animal use of structures often increases with time as they become accustomed to using crossings, especially animals habituated to using wildlife crossings early in life.

The type of fencing is variable, usually 8 feet or greater in height for large carnivores like wolves, bears, and cougar. Aprons attached to the bottom of fencing may be required to keep animals from digging under, and electrical or barbed wire may be needed on top to minimize animals climbing over. Carnivores are very good diggers and climbers, factors that can frustrate efforts to force them into safe wildlife crossings. However, once accustomed to using wildlife crossings, most carnivores readily use them.

Acquisition of Key Wildlife Linkage Areas

While many of the core areas used by carnivores are public land, the lands that connect these areas are often private. This is true even in the West, where mountain and desert habitats are often public lands; however, the valley bottoms between these lands are usually private. There are a variety of means of securing private land linkages. Property can be purchased by wildlife agencies, or by land management, transportation, or conservation groups. Conservation easements can be bought or donated. In most situations, existing rural land management practices, particularly ranching or forestry, can be maintained. The primary concern is that key wildlife habitat linkages not be subdivided with small lots, housing developments, and associated factors such as roaming pets, unnatural food sources, increased traffic, and the general intolerance shown by humans toward many carnivores.

Securing key wildlife habitat linkage areas on private land is a complex, expensive, and sometimes controversial action. There is a large and expanding need for nonprofit groups familiar with easements and acquisitions to help negotiate wildlife habitat linkages through private lands. Government agencies often have a number of restrictions that may complicate or extend the time needed to take action on private lands when landowners have a desire or need for a fast sale. Budgets, policy, red tape, and politics can result in government-run land transactions taking too long and costing too much. These factors can be frustrating to private landowners, particularly those that have an immediate need to secure funds for taxes or other inheritance issues.

Transportation agencies often must deal with securing right-of-ways and mitigations for highway projects. Unbeknownst to most people, highway agencies often have large land easement and acquisition programs and are key entities in wetland protection and acquisition, riparian habitat management and acquisition, and the purchasing of open space, often with a combination of funding sources. An opportunity exists to combine highway mitigation funding, land management and wildlife agency funding, and funding from private nonprofit conservation groups (such as The Nature Conservancy, Rocky Mountain Elk Foundation, Ducks Unlimited, and Trout Unlimited) to secure large areas critical to wildlife and fish conservation. Such partnerships could allow large-sized projects not possible for any single entity. The potential of private/government conservation programs, coupled with transportation agencies' relatively large funding of required highway mitigation programs, is perhaps the greatest present-day opportunity for benefiting fish and wildlife conservation. Capturing this opportunity will take foresight, ingenuity, and innovation by agencies, public officials, and conservation leaders.

Developing and Implementing Clear Agency Policies

There is nothing more important to future carnivores and other wildlife than focusing on the combined forces of habitat loss and fragmentation. The amount and abundance of wildlife available to future generations will be a function of (1) the amount of habitat society provides for wildlife conservation; (2) the quality of that habitat (not relegating wildlife to just the marginal scrublands, rock, and ice); (3) the connectedness of these lands;

and (4) the allowance of natural processes such as fire and floods to exert their influence on the landscape. These factors define a functioning ecosystem. Carnivores are vital parts of ecosystems and one of the most sensitive to disruptions.

To ensure a future for carnivores and other wildlife populations, state and federal governments need clear policies stating that maintaining habitat connectivity is an important ecosystem objective and that highways and other developments of the land will consider these factors. Addressing habitat fragmentation is an interagency issue that involves highway, land management, and wildlife agencies at a minimum. National and state forests and parks cannot sustain future wildlife unless there is suitable connectivity of these and other key conservation lands. There are very few, if any, national forests, national wildlife refuges, or national parks that can sustain native carnivores and other wide-ranging species within their boundaries. To ensure that these species persist, these lands must be interconnected into a matrix. The likelihood of this happening without a habitat connectivity and restoration policy is dim at best. The Great Plains, upper Midwest, Northeast, and Southeast are in a crisis situation, with little time left to develop effective habitat connectivity plans and programs. Wildlife habitat linkage analysis is needed in all regions of the United States, Canada, and Mexico.

Transportation agencies are probably the best source of funding for regional wildlife linkage analysis. Highways have the greatest impact on habitat fragmentation and urban sprawl. They also have the greatest funding resources. Without clear policy to address these issues, however, progress will be hodgepodge or nonexistent. At this time, there is no program to look at regional areas of the United States, Canada, and Mexico for the purpose of defining habitat connectivity.

Conclusion

Future highway developments offer both huge potential impacts to carnivores, as well as huge opportunities to contribute toward their conservation. Agencies and society have the technology to address these impacts. And we have the funding resources. Loss of habitat, habitat fragmentation, and highway-caused wildlife mortality can be reduced or eliminated. Progress in addressing these issues has been profound in the last decade, but so has the pace of highway development and human sprawl. Are we up to this challenge? We certainly are, but we will need to work closer together, to plan our highways and other developments better, and to integrate private and various government efforts. As Teddy Roosevelt once said, "We are not building our country for today—it is to last through the ages." Well said, Teddy.

Literature Cited

Bank, F. G., C. L. Irwin, G. L. Evink, M. E. Gray, S. Hagood, J. R. Kinar, A. Levy, D. Paulson, B. Ruediger, and R. M. Sauvajot. 2002. *Wildlife Habitat Connectivity Across European Highways*. Office of International Programs, U.S. Department of Transportation, Washington DC. Report # FHWA-PL-02-011.

Boecklen, W. J., and N. J. Gotelli. 1984. "Island biogeographic theory and conservation practice: Species-area or specious-area relationship?" *Biological Conservation* 29:63–80.

Brocke, R. H., K. A. Gustafson, and L. B. Fox. 1991. "Restoration of large predators: Potentials and problems," in *Challenges in the Conservation of Biological Resources: A Practitioner's Guide*, ed. D. J. Decker, M. E. Krasny, G. R. Groff, C. R. Smith, and D. W. Gross, 303–315. Westview Press, Boulder, CO.

Caroll, C., R. F. Noss, and P. C. Paquet. 2000. *Carnivores as Focal Species for Conservation Planning in the Rocky Mountain Region*. World Wildlife Fund Canada, Toronto.

———. 2000. "Factors influencing the effectiveness of wildlife underpasses in Banff National Park, Alberta, Canada," *Conservation Biology* 14:47–56.

Clevenger, A. P., and N. Waltho. 1999. "Dry drainage culvert use and design considerations for small and medium sized mammal movement across a major transportation corridor," in *Proceedings of the Third International Conference on Wildlife Ecology and Transportation*, ed. G. L. Evink, P. Garrett, and D. Zeigler, 263–287. FL-ER-73-99. Florida Department of Transportation, Tallahassee.

Colorado Division of Wildlife. 2000. "New satellite collars aid Division of Wildlife researchers in monitoring lynx in Colorado," *Wildlife Report,* June 29. Available at http://wildlife.state.co.us/cdnr_news/wildlife/2000629172048.html.

Copeland, J. 1994. Presentation at 1994 Western Forest Carnivore Committee meeting. Idaho Fish and Game Department, Idaho Falls.

Copeland, J. 1996. "Biology of the Wolverine in Central Idaho." Masters Thesis, University of Idaho.

Cox, J. A., and R. S. Kautz. 2000. *Habitat Conservation Needs of Rare and Imperiled Wildlife in Florida*. Office of Environmental Services, Florida Fish and Wildlife Commission, Tallahassee.

Ferreras, P., J. J. Aldama, J. F. Beltran, and M. Delibes. 1992. "Rates and causes of mortality in a fragmented population of Iberian lynx, *Felis pardina,* Temminck, 1824," *Biological Conservation* 61:197–202.

Forman, R. T. T., D. Sperling, J. A. Bissonette, A. P. Clevenger, C. D. Cutshell, V. H. Dale, L. Fahrig, R. France, C. R. Goldman, K. Heanue, J. A. Jones, F. J. Swanson, T. Turrentine, and T. C. Winter. 2003. *Road Ecology: Science and Solutions*. Island Press, Washington, DC.

Gibeau, M. L. 1993. "Use of urban habitat by coyotes in the vicinity of Banff, Alberta." M.S. thesis, University of Montana, Missoula.

Gibeau, M. L., and K. Heuer. 1996. "Effects of transportation corridors on large carnivores in the Bow River Valley, Alberta," in *Trends in Addressing Transportation Related Wildlife Mortality,* ed. G. L. Evink, P. Garrett, D. Ziegler, and J. Berry. FL-ER-58-96. State of Florida Department of Transportation, Tallahassee.

Gilbert, T., and J. Wooding. 1996. "An overview of black bear roadkills in Florida, 1976–1995," in *Trends in Addressing Transportation Related Wildlife Mortality,* ed. G. L. Evink, P. Garrett, D. Ziegler, and J. Berry. FL-ER-58-96. State of Florida Department of Transportation, Tallahassee.

Harris, L. D., and P. B. Gallagher. 1989. "New initiatives for wildlife conservation: The need for movement corridors," in *Preserving Communities and Corridors,* 11–34. Defenders of Wildlife: Washington, DC.

Harris, L. D., C. Maser, and A. McKee. 1982. "Patterns of old growth harvest and implications for Cascades wildlife," *Transactions of the North American Wildlife and Natural Resources Conference* 47:374–392.

Jalkotzy, M. G., P. I. Ross, and M. D. Nasserden. 1997. *The Effects of Linear Development on Wildlife: A Review of Selected Scientific Literature.* Prepared for the Canadian Association of Petroleum Producers, Calgary, AB.

Jenkins, K. 1996. "Texas Department of Transportation wildlife activities," in *Trends in Addressing Transportation Related Wildlife Mortality,* ed. G. L. Evink, P. Garrett, D. Ziegler, and J. Berry. FL-ER-58-96. State of Florida Department of Transportation, Tallahassee.

Koehler, G. M., and J. D. Brittell. 1990. "Managing spruce-fir habitat for lynx and snowshoe hares," *Journal of Forestry* (October):10–14.

Krohn, W. B., S. M. Arthur, and T. F. Paragi. 1994. "Mortality and vulnerability of a heavily trapped fisher population," in *Martens, Sables, and Fishers: Biology and Conservation,* ed. S. W. Buskirk, A. S. Harested, M. G. Raphael, and R. A. Powell, 137–145. Cornell University Press, Ithaca, NY.

Lewis, D. 1996. Presentation at 1996 Western Forest Carnivore Committee meeting, Canmore, British Columbia.

Maehr, D. S. 1984. "Animal habitat isolation by roads and agricultural fields," *Biological Conservation* 29:81–96.

Maehr, D. S., E. D. Land, and M. E. Roelke. 1991. "Mortality pattern of panthers in Southwest Florida," *Proceedings of the Annual Conference of Southeast Association of Fish and Wildlife Agencies* 45:201–207.

Matthiae, P., and F. Stearns. 1981. "Mammals in forest islands in southeast Wisconsin," in *Forest Island Dynamics in Man Dominated Landscapes,* ed. R. Burgess and D. Sharpe. Springer Verlag, New York.

Mattson, D. J., R. Knight, and B. Blanchard. 1987. "The effects of developments and primary road systems on grizzly bear habitat use in Yellowstone National Park, Wyoming," *International Conference on Bear Research and Management* 7:259–273.

Mech, L. D., and L. Boitani. 2003. "Wolf Social Ecology," in *Wolves: Behavior, Ecology, and Conservation* ed. L. D. Mech and L. Boitani, pp 1–34. University of Chicago Press, Chicago.

Noss, R. F. 1991. "Landscape connectivity: Different functions at different scales," in *Landscape Linkages and Biodiversity,* ed. W. E. Hudson, 27–39. Island Press, Washington, DC.

Paquet, P. C. 1995. *Large Carnivore Conservation in the Rocky Mountains.* World Wildlife Fund Canada, Toronto.

Paquet, P. C., and C. Callaghan. 1996. "Effects of linear developments on winter movements of gray wolves in the Bow River Valley of Banff National Park, Alberta," in *Trends in Addressing*

Transportation Related Wildlife Mortality, ed. G. L. Evink, P. Garrett, D. Ziegler, and J. Berry. FL-ER-58-96. State of Florida Department of Transportation, Tallahassee.

Reed, R. A., J. Johnson-Barnard, and W. L. Baker. 1996. "Contribution of roads to forest fragmentation in the Rocky Mountains," *Conservation Biology* 10:1098–1106.

Ruediger, B. 1996. "The relationship between rare carnivores and highways," in *Trends in Addressing Transportation Related Wildlife Mortality,* ed. G. L. Evink, P. Garrett, D. Ziegler, and J. Berry. FL-ER-58-96. State of Florida Department of Transportation, Tallahassee.

Ruediger, B. 2001. *Draft Report to the Interagency Grizzly Bear Working Group on Wildlife Linkage Habitat.* Available at: www.fs.fed.us/r1/wildlife/igbc/Linkage/LinkageReport.htm.

Ruediger, B., J. Claar, and J. Gore. 1999. "Restoration of Carnivore Habitat Connectivity in the Northern Rocky Mountains," in *Proceedings of the Third International Conference on Wildlife Ecology and Transportation,* ed G. L. Evink, G.L., P. Garrett, and D. Zeigler. FL-ER-73-99, Florida Department of Transportation, Tallahassee.

Ruggiero, L. F., K. Aubry, S. W. Buskirk, G. M. Koeler, C. J. Krebs, K. S. McKelvey, and J. R. Squires. 2000. *Ecology and Conservation of Lynx in the United States.* University of Colorado Press, Boulder.

Ruggiero, L. F., K. B. Aubry, S. W. Buskirk, L. J. Lyon, and W. J. Zielinski. 1994. *The Scientific Basis for Conserving Forest Carnivores: American Marten, Fisher, Lynx, and Wolverine in the Western United States.* USDA Forest Service General Technical Report, RM-254. Rocky Mountain Forest and Range Experiment Station, Fort Collins, CO.

Servheen, C., J. S. Waller, and P. Sandstrom. 2001. *Identification and Management of Linkage Zones for Grizzly Bears Between Large Blocks of Public Land in the Northern Rocky Mountains.* U.S. Fish and Wildlife Service and University of Montana, Missoula, MT.

Singleton, P. H., W. L. Gaines, and J. F. Lehmkuhl. 2002. *Landscape Permeability for Large Carnivores in Washington: A Geographic Information System Weighted-Distance and Least-Cost Corridor Assessment.* Research Paper PNW-RP-540. USDA Forest Service Pacific Northwest Research Station, Portland, OR.

Walker, R., and L. Craighead. 1997. "Analyzing wildlife movement corridors in Montana using GIS," in *Proceedings of the 1997 Environmental Sciences Research Institute International User Conference, Redlands, CA.* Environmental Sciences Research Institute. Walker and Craighead paper available at: http://gis.esri.com/library/userconf/proc97/proc97/to150/pap116/p116.htm.

Weaver, J. L., P. C. Paquet, and L. F. Ruggiero. 1996. "Resilience and conservation of large carnivores in the Rocky Mountains," *Conservation Biology* 10:964–976.

Wilcox, B. A., and D. D. Murphy. 1985. "Conservation strategy: The effects of fragmentation on extinction," *American Naturalist* 125: 879–887.

CHAPTER 8

Living with Fierce Creatures? An Overview and Models of Mammalian Carnivore Conservation

David J. Mattson

Many mammalian carnivores have been lost during the last 200 years. Tigers (*Panthera tigris*) have been extirpated from much of their range, as have lions (*P. leo*), leopards (*P. pardus*), and cheetahs (*Acinonyx jubatus*) from most of Asia. Sun (*Helarctos malayensis*), sloth (*Melursus ursinus*), and spectacled bears (*Tremarctos ornatus*) are reduced to relict populations. Most small felids of the Indian subcontinent are threatened with extirpation. Even circumboreal species such as the brown bear (*Ursus arctos*) and wolf (*Canis lupus*) have experienced major range contractions, to the point of endangerment in Europe and the contiguous United States. In fact, 32% of all carnivore species are currently at risk (Ceballos and Brown 1995). On the other hand, some carnivores have fared well. Coyotes (*C. latrans*), jackals (*C. aureus*), and red foxes (*Vulpes vulpes*) have multiplied, and skunks (family Mephitidae) and weasels (*Mustela* spp.) flourish except in the most heavily human-impacted environments.

Humans have been and continue to be the cause of most carnivore losses. Human impacts have been documented extensively (e.g., IUCN Cat Specialist Group; IUCN Canid Specialist Group). Carnivores are killed legally and illegally for a myriad of reasons, including the belief that they are a threat to human safety, depredators of livestock, and depredators of wild game. They are also killed for their pelts, for other valued body parts, for prestige, and out of animosity and intolerance. Carnivores also die because their prey are decimated by overhunting or because the habitats that support them are destroyed. Yet carnivores are often revered, and are widely represented in art, stories, and religious motifs. The giant panda

(*Ailuropoda melanoleuca*) elicits sympathy and financial outpourings worldwide, enough to warrant its adoption as a symbol of the World Wildlife Fund. Even tigers and wolves have their human enthusiasts, perhaps in numbers greater worldwide than their detractors.

Humans have clearly impacted some carnivores more than others. This begs a number of questions: What human perspectives and behaviors allow us to live with carnivores? Do humans create irreducible impacts and, if so, to what extent and under what circumstances? What traits make carnivores intrinsically less resilient than other taxa to persecution and conflict with humans? What landscape features exacerbate this conflict? Answering these questions will be important to gaining the insight we need to live with carnivores, but contingent on organizing lessons and identifying key drivers at both coarse and fine scales.

In this chapter I present general models of factors that govern the fates of carnivores. I emphasize broad-scale patterns and concepts, particularly as they relate to human-carnivore interactions. I focus on biophysical elements, realizing that the policy process is critically important but beyond the scope of this chapter. I also focus on conservation rather than on harvestable surpluses, the traditional topic of game management. For many rare and vanishing carnivores, more is required of us than simply adjusting trapping or hunting regulations.

The Limits of Island Biogeography

Much of the past research explaining mammalian extirpations has adopted the paradigm and associated models of island biogeography. The size of protected areas, which function as islands in a sea of developed area, is invoked to explain the richness of terrestrial vertebrate faunas. Worldwide, relatively strong relationships have been shown between species richness or losses and sizes of national parks and other nature reserves (e.g., Woodroffe and Ginsberg 1998). Some analyses have included the often substantial effects of local human densities and population growth rates (e.g., Bashares et al. 2001), suggesting that the effectiveness of reserves is impacted by the human matrix, primarily through intrusions along edges.

This focus on protected area size has limitations, especially for generating the insight needed to guide conservation action. For one, protected area size explains typically less than half of species losses or richness. Second, "protected area" is merely shorthand for a policy prescription governing a range of elements, including human settlement, the carrying of firearms, the presence of livestock, the extraction of natural resources, and wildlife protection. In reality, prescriptions for protected areas vary widely. Some allow settlement, some allow grazing, and some allow subsistence hunting. Third, whatever the prescription, levels of enforcement and implementation differ, to the extent that some "protected areas," as drawn on maps, offer virtually no protection at all. Last, even naïvely assuming that "protected areas" are all functionally the same, reliance on reserves as a prescriptive response to endangerment holds little prospect of saving most species. No single protected area is large enough to ensure the future of any species, nor are there many opportunities left to set aside large areas for wildlife conservation excluding most humans and human activities.

We need robust models of the causes of endangerment to address the challenges of carnivore conservation. Such models can aid the design and evaluation of conservation efforts, focus research efforts, and provide a language for communication. To serve these ends, the rest of this chapter is devoted to developing functional models that contain more than just protected area size and human density, and that distinguish human-related factors from biological ones.

Human Factors Governing Endangerment

Human factors that more immediately impact carnivores are fruitfully distinguished from those that operate through intervening effects; that is, "proximal factors" can be distinguished from "distal factors." Proximal factors operate closer in time and space to outcomes such as carnivore births or deaths. Distal factors (e.g., road densities or broad-scale market conditions) are conditioning, in that they preconfigure the frequency and intensity of proximal effects.

Proximal Factors

Proximal factors constitute direct sources of mortality. Proximal factors that endanger carnivores include harvest of body pats, retaliation for depredation, collisions with vehicles, loss of prey, loss of habitat, and disease.

HARVEST OF BODY PARTS

Humans commonly kill carnivores because they have economically valuable body parts. Most often this is the pelt, desired as adornment or functional clothing. Less often organs or bones are sought, as with tiger bones and bear gall bladders used for medicinal ingredients. The worldwide role of fur harvest in decimating populations of felids and mustelids is pervasive. Fur harvest devastated Eurasian lynx (*Lynx lynx*) and Canada lynx (*L. canadensis*), otter (*Lutra lutra* and *L. canadensis*), marten (*Martes martes* and *M. americana*), mink (*Mustela lutreola*), and fisher (*Martes pennanti*) populations in Europe and the contiguous United States during the 1800s and early 1900s (Mason and Macdonald 1986; Griffiths 2000; IUCN Cat Specialist Group). Fur harvest continues to severely threaten felids of South America and Asia, including snow (*Uncia uncia*) and clouded (*Neofelis nebulosa*) leopards (IUCN Cat Specialist Group). Fortunately, this kind of threat can be relatively easily addressed. Although complicated by distal factors (see below), regulations can target the specific act of harvest to either achieve sustainable levels or curtail it altogether. Such a response was responsible for the recovery of many furbearers, especially in Europe and North America.

RETALIATION FOR DEPREDATION

Carnivores are killed virtually everywhere to retaliate for, or prevent depredation of, livestock. Depredation control has impacted and even threatened populations of species as diverse as polecats (*Mustela putorius*) in Europe, snow leopards in India, pumas (*Puma concolor*) in the United States, jackals in the Middle East, and lynx and wolverine (*Gulo gulo*) in Scandinavia. Worldwide, sheep and goats, in contrast to cattle, horses, and yaks, are the most vulnerable to predation, primarily because of their small size and often unattended grazing (Mattson 1990; Jackson et al. 1996; Landa et al. 1999; Logan and Sweanor 2000).

Depredation is typically the outcome of a syndrome involving eradication of native prey, introduction of vulnerable livestock, and elimination of dominant predators. The densities and relative proportions of native prey and livestock have major effects on depredation (Jackson et al. 1996; Schaller 1998). Wolves, pumas, and grizzly bears (*Ursus arctos horribilis*) were slaughtered in the Rocky Mountains during the late 1800s at a time when native ungulates were virtually eliminated by subsistence and market hunting and replaced by a flood of sheep and cattle (Brown 1983, 1996). Depredations predictably mounted, along with retaliatory campaigns to extirpate all predators. Similar scenarios persist in such places as Tibet and Mongolia (Reading et al. 1998; Schaller 1998).

Secondary to this effect, depredations can escalate as medium-sized carnivores kept in check by larger predators become more numerous when the larger predators are removed (Terborgh et al. 1999). High levels of depredation on sheep by coyotes in the Rocky Mountains can be partly explained by the elimination of wolves, which are known to keep coyote populations in check (Crabtree and Sheldon 1999). There is similar evidence from bears and pumas that local removals of dominant individuals to protect livestock can actually increase losses as dispersing juveniles fill the vacant niche in higher densities than before and readily turn to killing sheep or raiding beehives (Mattson 1990; Logan and Sweanor 2000).

The most common response to depredation has been generalized persecution of predators. Obviously, if carried on over a large enough area, employing enough resources, and for a long enough time, all creatures that potentially eat livestock can be eradicated. Such a solution has been achieved de facto or by design in many areas. However, the obvious problems with such an approach are that it threatens carnivore species with extinction and often unravels ecological webs. Explosions of rodent and lagomorph populations are prime examples of unforeseen consequences. Killing predators piecemeal in retaliation for depredations has only limited short-term efficacy, contributes little to long-term solutions, and may even be counterproductive (e.g., Stahl et al. 2001). Losses typically change little or may even increase. Solutions with better long-term prospects include reducing livestock numbers, raising species that are less vulnerable to predation (e.g., cattle or yaks), improving husbandry, and increasing the abundance of native prey.

COLLISION WITH MOTOR VEHICLES

Collisions with vehicles traveling at high speeds can be a major cause of mortality where there are numerous heavily trafficked roads. Under such circumstances all carnivore species are impacted, and some even endangered. Because high densities of well-maintained roads are emblematic of affluence, populations endangered by roadkills are primarily restricted to Europe and North America, including the panther (*Puma concolor coryi*) and black bear (*Ursus americanus floridanus*) in Florida, the badger (*Meles meles*) in northern Europe, and the Pardel lynx (*Lynx pardinus*) in southern Spain (Ferreras et al. 1992; Forman and Alexander 1998). High-speed roads can also have substantial local effects, as on wolves and bears in Slovenia and the Rocky Mountains, where high-relief mountain valleys often contain major highways that transect natural travel routes and attract carnivores to scavenge roadkills or spilled edibles (Kaczensky 1996; Clevenger et al. 2001). Barring the removal of roads, erection of barriers along roadways has been the primary response to roadkills. Where this has fragmented populations, under- or overpasses have sometimes been installed to promote connectivity, with varying success.

LOSS OF PREY

Prey abundance and size largely determine birth rates and native densities of carnivores. Some carnivores have declined, or had their recovery forestalled, because of insufficient prey. This has most often been the case for small or medium-sized predators that specialize on few prey species. Canada lynx specializing on forest-dwelling hares, *Lepus americanus*) and black-footed ferrets (*Mustela nigripes;* specializing on prairie dogs, *Cynomys* spp.) are prominent examples (Miller et al. 1996; Ruggiero et al. 1999). Loss of prey caused by deforestation is also suspected of contributing to the decline of forest-dwelling carnivores such as Asia's clouded leopard and marbled (*Pardofelis marmorata*) and golden (*Catopuma temminckii*) cats, South America's oncilla (*Leopardus tigrinus*), and martens of Europe and North America (Griffiths 2000; IUCN Cat Specialist Group). Otters and mink have also been impacted by loss of aquatic prey, most often in industrialized countries with unchecked water pollution (Mason and MacDonald 1986).

Remediation for habitat and prey loss is often difficult because causes are rooted in broad-scale human economic and demographic forces (see distal factors below). However, otters have recovered with control of water pollution, as have forest-dwelling mustelids and felids where timber harvest and deforestation have been curbed. Even so, in all cases protection of the carnivores themselves has also been required.

LOSS OF COVER OR HUNTING HABITAT

Loss of "cover" and "habitat" is often invoked to explain carnivore declines. However, in most cases, functional connections to thermal regulation, security, prey abundance, or favorable hunting conditions are not known. In North America, martens and fishers have been associated with forest cover per se. However this connection seems to be largely an artifact of dependence on forest-associated prey such as red squirrels (*Tamiasciurus* spp.), porcupines (*Erethizon dorsatum*), and hares, although martens seem to be more vulnerable to avian predators near forest edges (e.g., Hargis et al. 1999). Wolves are commonly associated with forest cover in Europe, apparently independent of prey and human densities (e.g., Massolo and Meriggi 1998). But wolves currently occupy tundra and were once abundant in nonforest environments at mid to low latitudes. In contrast to losses of forest-dependent prey, the effects of losing forest cover per se are typically poorly understood, despite the fact that these effects are commonly invoked.

DISEASE

Disease can be a major threat to carnivores, especially canids. The geographic scope of this threat is worldwide. Ethiopian wolves (*Canis simensis*) and African wild dogs *(Lycaon pictus)* are killed by rabies; Mednyi Island arctic foxes (*Alopex lagopus semenovi*) by mange; bat-eared foxes (*Otocyon megaiotis*) of the Serengeti by distemper; and crab-eating foxes (*Cerocyon thous*) of Brazil by leishmaniasis (e.g., Murray et al. 1999). Domestic dogs play a major role in harboring and spreading rabies and distemper. However, the threat of disease is not restricted to the dog family. Distemper nearly killed the last black-footed ferrets (Miller et al. 1996). More recently, distemper has been epidemic among African lions (*Panthera Leo;* Roelke-Parker et al. 1996). One of the more disturbing aspects of disease is its unpredictable nature, both in the outbreak of known pathogens and in the evolution of more lethal or previously unknown strains. Control of disease can be costly, and usually involves inoculating or treating not only affected carnivores but also other known vectors, including domesticated animals.

Distal Factors

Distal factors are those features of the landscape and human community that increase the likelihood of carnivore–human conflict. Distal factors include road densities, human densities, wealth, overlap in concentration, and values and perspectives.

ROAD DENSITIES

The negative effects of roads on wildlife are legion. Carnivores such as wolves, wolverines, black bears, grizzly bears, and bobcats (*Lynx rufus*) either avoid crossing highways or are found only in areas with few or no roads (e.g., Mattson et al. 1996a; Carroll et al. 2001). Roads allow given numbers of humans to use a landscape more intensively than would otherwise be possible. This matters because most carnivores are killed by humans near roads. As road densities increase, demographic source areas for vulnerable carnivores disappear. As described above, vehicles on roads are also a cause of death, in addition to road-bound armed humans. Perhaps most important, roads facilitate the settlement of previously unsettled areas and, with that, deforestation, expansion of croplands, and increases in livestock, with related impacts on native prey and conflicts with carnivores (see proximal factors above). Roads also often cause increased pollution of waterways, especially in mountainous areas, with related effects on aquatic organisms and carnivores such as otters and mink that prey on them. As a rule, roads are closely associated with a host of human activities that have been shown to endanger animals and, in fact, largely govern their spread and intensity (Trombulak and Frissell 2000). That said, roadbeds themselves are rarely the direct cause of problems. Rather, it is the activities and behaviors of the humans that use them.

HUMAN DENSITIES

All else equal, human densities also affect where carnivores survive. As mentioned above, carnivores have more often been extirpated in protected areas surrounded by high rather than low densities of humans, especially where human populations were growing rapidly. Italian wolves, (*C.l. lupus* grizzly and brown bears, and Siberian tigers (*P.t. altaica*) have persisted only where there are few humans, typically <1–8 people/km^2 (Massolo and Meriggi 1998; Miquelle et al. 1999; Mattson and Merrill 2002). Reflecting these negative associations, more mammal species are endangered, globally, in countries where there are high versus low human densities (e.g., Kerr and Currie 1995). Large numbers of humans are problematic for obvious reasons. Where there are more people, carnivores are more likely to encounter them, often with negative results. There are also likely to be more livestock, more croplands, more industrial activities, and more pollution, all with predictably negative impacts on most carnivores.

Mountain areas highlight the importance of human numbers to carnivore conservation. Mountains are often seen as refuges for wildlife. However, this is primarily an artifact of low human density. Steep terrain complicates the construction of homes, roads, irrigation systems, and agricultural plots. Moreover, the thin soils and short growing seasons of high-latitude mountains make them unproductive environments, despite often adequate precipitation. It is thus not surprising that a recent appraisal of ecological integrity in the Columbia River basin of the United States found moderate to high levels of this elusive quality only in mountainous areas.

Arid equatorial mountains emphasize the extent to which mountain sanctuaries are contingent on an absence of humans. In the central Andes of South America and in the Abyssinian Highlands of Africa, humans are concentrated in mountainous highlands, primarily either because of aridity or because of the prevalence of disease at lower elevations. In Bolivia, Ecuador, and Peru, human densities in more arable parts of the Altiplano reach 16–70 people/km^2 (Medina). Similarly, at 50 people/km^2, the mountains of Ethiopia are the most densely populated agricultural region in Africa (Funnell and Parish 2001). Not surprisingly, wildlife populations have suffered in these heavily populated mountain areas, including the Ethiopian wolf (Gottelli and Sillero-Zubiri 1992) and Andean mountain cat (*Oreailurus jacobitus*) (IUCN Cat Specialist Group).

WEALTH

The relative wealth of individuals and nations has mixed, but substantial, effects on carnivores. Proximally, where people living near carnivores are poor, there are often greater incentives to kill carnivores for profit or to retaliate for lost livestock. Even a few lost livestock can cost as much as 25%–50% of an annual disposable income for subsistence farmers in large parts of Asia and Africa (e.g., Oli et al. 1994). Similarly, the value of a snow leopard pelt can equal as much as 10% of the annual income of a rural resident of the Himalayas (Fox 1991), enough to perhaps risk running afoul of protective regulations. Where protected areas harbor forests and forest-associated carnivores, poverty can lead rural residents who are desperate for fuel or new farm plots to defy protective regulations and deforest large areas, with negative consequences for the carnivores. Impoverished people worldwide also commonly depend on wild game to augment protein in otherwise protein-deficient diets, with resulting depletion of herbivore populations that would otherwise support native carnivores (e.g., Rao and Gowan 2002). At broader scales, poverty is usually associated with social inequities and poor governance. These factors often lead to corruption, civil strife, and the breakdown of civil order, all of which impair the implementation of regulations designed to protect wildlife, including wildlife in ostensibly protected areas (e.g., Dudley et al. 2002).

On the other hand, wealth breeds its own set of problems for carnivores.

Increasing affluence in cultures that traditionally value furs or other carnivore body parts can cause increased demand for such things, with resulting increased exploitation of carnivores in countries as much as half a world away. Such has been the outcome of increasing affluence in China and Southeast Asia, with dire consequences for tigers and bears (e.g., Seidensticker 1997). Initial stages of industrialization also can fuel increased deforestation as well as increased conversion of native vegetation to cropland. World-record conversions of native grasslands and forests to agriculture occurred in the former Soviet Union during modernizations of the 1950s and 1960s (Houghton 1994). Industrialization can increase air and water pollution to levels at which fish populations are decimated and forests killed. However, as industrialization progresses, there is a trend toward increased forest and wildlife conservation and better control of pollution, because of changing values and product substitution from foreign sources or through technological innovation (e.g., Czech et al. 2000). However, at this level of development, high-speed roads can be so extensive and ownership of cars so common that roadkill becomes a major problem.

JOINT CONCENTRATIONS

The extent to which humans and carnivores are jointly concentrated in a landscape can have major effects on the persistence of carnivore populations. If settlements, roads, livestock, and people are concentrated in the same habitats as predators and their prey, then humans will logically have a much greater impact than if spatial arrangements are otherwise. Despite its potential importance, this landscape-level phenomenon has rarely been investigated, probably because the spatial scope of required data is prohibitive. However, the fates of wolves in Spain and grizzly bears in the contiguous United States provide evidence for the potential importance of this factor. In Spain, wolves have fared better in the northern mountains, where humans are concentrated in valley bottoms and away from most big game, as compared with wolves in the northwestern mountains, where villages and people are more dispersed (Reig et al. 1992).

Likewise, grizzly bears were extirpated more quickly between 1850 and 1920 in the contiguous United States, where they relied on foods that brought them into frequent contact with humans (i.e., spawning salmonids and bison [*Bison bison*]), as compared with where important foods kept them dispersed and at higher elevations (Mattson and Merrill 2002). Grizzly bears in the Yellowstone region continue to be affected in this way (Pease and Mattson 1999). When high-elevation crops of whitebark pine (*Pinus albicaulis*) seeds are abundant, grizzly bears concentrate away from people, and few die. When seeds are scarce, the bears spend much more time near people, and death rates increase.

Mountain areas highlight the importance of the concentration factor. High levels of vertical variation in climate, soils, and vegetation typify mountains. Biological productivity is correspondingly concentrated in zonal bands determined by regional climatic conditions. For obvious reasons, humans tend to concentrate in the more productive flatter parts of mountains, most often in valley bottoms. As a result, macroscale patterns of human settlement and activity are often spatially more variable in mountains as compared with flatlands. For carnivores that concentrate in the same kinds of mountain places as humans, this heterogeneity can heighten conflict and vulnerability to human exploitation. For carnivores that favor zones unused by humans, this heterogeneity can work in the carnivores' favor.

High-speed highways and railroads pose a special problem for wildlife in mountains, primarily because of the concentration factor. Most transportation routes of this sort follow valley bottoms. As a consequence, they transect not only concentrations of potential prey, but also major cross-valley travel routes. By circumstance and constraints on design, mountain highways often seem built so as to maximize the odds that wildlife will be placed in the path of speeding vehicles, with potentially catastrophic consequences. For carnivores, risks are aggravated by their tendency to scavenge roadside road-killed animals. For omnivores such as bears, attractants can include grain spilled from passing trains or semitrailers, or even clover planted to stabilize roadside ditches.

VALUES AND PERSPECTIVES

It seems self-evident that human perspectives about carnivores would have major effects on the rates at which we kill them. This factor logically governs the extent to which we seek carnivores out with lethal intent, or how we otherwise respond to an encounter once it occurs. Culturally inculcated values and related myth systems largely determine perspectives on nature and wildlife. For this reason, perspectives vary substantially among cultures, with time, and even among regions.

In the United States, residents of interior regions, where most remaining large carnivores survive, tend to express negative environmental attitudes more often than do residents of coastal regions, where most large carnivores have been extirpated (e.g., Hall and Kerr 1991). However, the most pronounced differences in perspectives are among or between age groups, employment sectors, genders, and those with advanced education versus those without (e.g., Kellert 1985). Those who are better educated, younger, female, and engaged in nonagricultural employment are generally more accepting of carnivores than those who are otherwise.

Although attitudes and perspectives about carnivores are relatively well documented, little has been done to show explicitly how these outlooks translate into action. Intuitively, relations between attitudes and a willingness to kill carnivores seem self-evident. However, empirical observations substantiating the nature and magnitude of such effects are virtually nonexistent.

Biological Factors Governing Endangerment

Compared with human factors, there are fewer factors in the biological realm that drive carnivore conservation, most prominently body size, range size, and degree of prey specialization.

Body Size

Carnivore size is perhaps the single most important biological factor governing the nature of interactions with humans and related levels of endangerment. Body size largely determines reproductive rate (Hennemann 1983); longevity (Wootton 1987); scale of movements (Gittleman and Harvey 1982); and size and diversity of prey (Gittleman 1985). As size increases so do life span, range size, and prey size, while reproductive rates drop. As prey size increases, to the point where sheep, cattle, yaks, and horses are included, so does the per capita monetary loss entailed by depredation. Moreover, large carnivores such as lions, tigers, leopards, and bears are more likely to view humans as potential prey, or to kill humans while defending themselves or their offspring. Thus, compared with small carnivores, large ones more frequently encounter humans because of their large individual ranges; are more often killed during an encounter because they are viewed as a threat or an unacceptable cost; and less often replace themselves because of a low birth rate.

All else equal, large carnivores are predictably more prone than small ones to endangerment by direct human-caused mortality. On the other hand, compared with large-bodied carnivores, small or medium-sized species are more often threatened by loss of prey, as such, because diet diversity tends to decline with body size (Gittleman 1985).

Range Size

Range size is not completely determined by body size. For a given body mass, different carnivore species or genders can use areas of substantially different size. This translates into substantially different odds of encountering humans or human features such as roads. Thus, range size per se explains extinctions from protected areas better than does body size (Woodroffe 2001). Wolves and wolverines, although small compared with black and grizzly bears, are probably at equal or greater risk of endangerment because of their comparatively more extensive movements (Weaver et al. 1996). Moreover, populations of given species living in highly productive environments are probably more robust to human stressors as compared with populations occupying impoverished habitats. For most carnivores, range sizes are smaller where habitat is more productive. This probably explains, in part, why populations of relatively sedentary European brown bears (*U.a. arctos*) and wolves (*C.l. signatus*) have been able to survive among densities of humans that have otherwise eliminated populations of wider-ranging conspecifics elsewhere.

Behavioral Flexibility

Compared with habitat and prey generalists, specialized predators are more vulnerable to human impacts. These types of predators have morphological or behavioral traits that make it difficult for them to switch prey or hunting habitats. Typically, specialized predators coevolved with prey that were potentially superabundant over large areas. As noted above, lynx and hares, and black-footed ferrets and prairie dogs, are classic examples of such pairings. Humans can catastrophically affect even superabundant prey. Such was the case for prairie dogs—and black-footed ferrets—in the United States during the late 1800s and early 1900s (Miller et al. 1996). Millions of prairie dogs were killed during massive extermination programs, with ferrets the ancillary victims. By contrast, species that are behaviorally and dietarily flexible often fare better in the face of human encroachment, sometimes despite intensive persecution. Coyotes and jackals are prime examples.

An Endangerment-Relevant Taxonomy

The preceding information implies a relatively simple taxonomy for carnivores that pertains to the likelihood and nature of endangerment posed by human activities:

> *Large carnivores* are prone to endangerment because they are wide ranging, often a threat to human safety and assets, and relatively unfecund. They die almost exclusively of human causes where they are in contact with people, and often at a rate that exceeds births.
>
> "*Generalists*" are the most robust of any carnivores to human impacts. Because they are smaller than "large carnivores," they are more fecund and so better able to compensate for high levels of mortality. They are also better able to switch prey and habitats as compared with the specialists and so can compensate for loss of certain prey species or hunting habitats to human encroachment. Many of these species also can subsist on human-related foods such as garbage.
>
> *Specialists,* like "large carnivores," also include species prone to endangerment, but more often because of prey shortages caused by human practices. Even so, populations of these carnivores are often depressed by direct exploitation, in addition to food shortages.

This taxonomy is confirmed by natural groupings of carnivore species from the central Rocky Mountains of North America. Figure 8.1 shows a cluster diagram for these species based on similarities of sensitivity to 25 different proximal limiting factors (Mattson et al., in review). Clustering was by average linkages using Euclidean distances. Grizzly bears, black bears, pumas, wolves, and wolverines are readily identifiable as the "large carnivores," dying primarily because humans kill them, and dependent on large-bodied ungulates (especially wapiti, *Cervus elaphus*) for food. The "generalists" comprise about half of the species, including canids, some mustelids, bobcats, and the single procyonid (raccoons, *Procyon lotor*). These species share sensitivity to targeted trapping and exhibit diverse diets. Not unexpectedly, "specialists" are "danglers" in the diagram. This follows from each having unique habitat or prey relations that entail unique limiting factors. Consequently, they don't group closely with other species. By this reckoning the "specialists" are otters, fishers, lynx, least weasels, and possibly martens, associated with fish, porcupines, hares, very small rodents, and red squirrels, respectively.

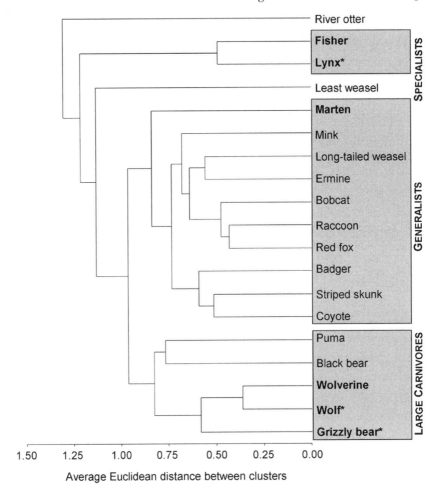

FIGURE 8.1 Cluster diagram of central Rocky Mountain carnivores based on average Euclidean distances and similarity of sensitivities to limiting factors. Bolded names indicate carnivores provided some level of protection under state or federal laws; names followed by an asterisk (*) are species protected under the U.S. Endangered Species Act.

Species that are endangered, threatened, or otherwise managed to promote conservation in the central Rocky Mountains are identified by bolding in Figure 8.1. Consistent with the expectations above, all of these special-status species are either large carnivores or prey specialists. As a basis for identifying "large carnivores" elsewhere, these results suggest that this category consists of species with adult females weighing >15 kg, or

>8 kg when range sizes are exceptionally large given the mass of the animal (e.g., wolverines). "Specialists" consist of species whose diets consist almost wholly of one or two prey items.

The effect of body mass and the demarcation of large carnivores are confirmed by patterns of endangerment among carnivore communities from four mountain environments: the central Rocky Mountains, the Andes, the southern mountains of Europe, and the Himalayas (Table 8.1). Treating different subspecies of the same species separately, 38% of the 74 carnivore taxa in these four regions are listed as threatened to some degree by the World Conservation Union (i.e., the IUCN Red List; IUCN 2003) or under the Convention on International Trade in Endangered Species (i.e., the CITES Appendices;UNEP-WCMC 2004). However, the odds of listing increase substantially with body mass; they are also higher, all else equal, in the Andes and Himalayas as compared with Europe and North America (Figure 8.2).

In the logit-based model describing this pattern, the coefficient for the natural log of body mass (kg + 1) is 0.74, and intercepts for the Andes, the Himalayas, Europe, and the Rocky Mountains are −0.75, −1.39, −2.15, and −2.28, respectively. Taxonomic family or genus had no effect. The model explains about 39% of variation in listings, and the statistical fit is acceptable (G^2 = 68, df = 64, and P = 0.35). Averaging over the four mountain environments, the probability of being listed equals 0.5 at a body mass of 8.2 kg, suggesting that this size distinguishes "large" from "small" carnivores, at least with respect to risk of endangerment. By this reckoning, 36% of the carnivores are "large," of which 48% are listed.

Two General Models of Endangerment

Large carnivores and specialists are the two groups most prone to endangerment by humans. By contrast, generalists are usually suited to management under the traditional paradigms of game animal and furbearer, with the proviso that even generalists can become endangered in the most human-impacted environments. Because large carnivores and specialists are typically endangered for different basic reasons, it is helpful to formulate models specific to each group. For large carnivores, the model logically focuses on factors driving or mediating direct human-caused mortality. For specialists, the model includes some of the same factors, but ancillary to a focus on factors driving loss of prey.

The fundamental outcome of interest for carnivores in the conceptual models shown in Figures 8.3 and 8.4 is finite rate of population change (λ). Lower growth rates equate to greater risks of endangerment. Growth rates are determined by differences between birth and death rates. Given a focus on human-caused mortality and the prevalence of this cause among carnivores in contact with people, death rate is governed in these models by how often a carnivore encounters people (i.e., "encounter rate"), and the probability that an encounter will turn lethal for the animal (i.e., "encounter deadly") (Mattson et al. 1996b). Factors associated with humans (boxes with rounded edges) are also distinguished from factors associated with carnivores (square boxes with solid delineations). Square boxes surrounded by dashed lines identify phenomena at the interface of humans and carnivores. Shaded boxes denote factors where there is potential to intervene in the system to promote conservation.

Direct Human-Caused Mortality

Figure 8.3 shows a conceptual model of relations among human and biological factors governing the endangerment of carnivores threatened primarily by direct human-caused deaths. This model is relevant to large-bodied carnivores that potentially pose a direct threat to humans or prey on large-bodied livestock.

In the upper quadrant, a subsystem is identified related to frequency of depredation, the size of livestock killed, and the relative monetary value of these losses. This subsystem is largely driven by predator size, husbandry practices, densities of livestock, rates at which humans kill native prey, range carrying capacity (K), and per capita human income.

Another model focal point pertains to the likelihood that a human will kill a carnivore during an encounter. This phenomenon is driven by whether the human is armed, the relative monetary losses caused by depredation, the relative value of the carnivore's pelt or other body parts, the perceived threat posed by the carnivore, and the existence value ascribed to the carnivore by the human. A factor not illustrated is fear of reprisal from authorities in cases in which the carnivore is protected by law.

Roadkill is represented by rates of encounters with vehicles. Levels of affluence, human and road densities, and the extent to which humans and roads are concentrated in prime carnivore habitat drive this rate. In general, human and road densities and their concentrations in places where carnivores are most active also drive encounter rates with humans.

TABLE 8.1

Taxa and their endangerment status from four mountain carnivore communities: the central Rocky Mountains of North America; the southern mountains of Europe; the Himalayas of India, Nepal, and Tibet; and the Andes of South America. Body mass is for an average adult female.
M. Silva and J. A. Downing, CRC Handbook of Mammalian Body Masses (CRC Press, Boca Raton, FL, 1995).

Common Name	Latin Name	Mass (kg)	CITES	IUCN	Common Name	Latin Name	Mass (kg)	CITES	IUCN
Central Rocky Mountains of North America					Mountains of Southern Europe				
Grizzly bear	*Ursus arctos*	129	—	—	Brown bear	*Ursus arctos*	158	CI[1]	—
Black bear	*U. americanus*	76	—	—	Wolf	*Canis lupus signatus*	37.0	CI	VU[2], LR[3]
Raccoon	*Procyon lotor*	7.0	—	—	Golden jackal	*C. aureus*	11.8	—	—
Gray wolf	*Canis lupus nubilis*	27.6	—	—	Red fox	*Vulpes vulpes vulpes*	7.8	—	—
Coyote	*C. latrans*	14.4	—	—	Lynx	*Lynx lynx*	24.8	—	—
Red fox	*Vulpes vulpes macroura*	4.6	—	—	Pardel lynx	*L. pardinus*	13.0	CI	EN[4]
Puma	*Puma concolor*	42.7	—	—	Wild cat	*Felis silvestris*	7.1	—	—
Bobcat	*Lynx rufus*	9.0	—	—	Wolverine	*Gulo gulo*	18.2	—	VU
Canada lynx	*L. canadensis*	8.6	—	—	European badger	*Meles meles*	13.0	—	—
Wolverine	*Gulo gulo*	8.9	—	VU	Otter	*Lutra lutra*	5.0	CI	VU
Northern river otter	*Lutra canadensis*	9.1	—	—	Pine marten	*Martes martes*	1.3	—	—
Badger	*Taxidaea taxus*	6.3	—	—	Beech marten	*M. foina*	1.8	—	—
Fisher	*Martes pennanti*	4.5	—	—	Polecat	*Mustela putorius*	0.7	—	—
Striped skunk	*Mephitis mephitis*	2.3	—	—	European mink	*M. lutreola*	0.7	—	—
American marten	*Martes americana*	0.6	—	—	Stoat	*M. erminea*	0.2	—	—
Mink	*Mustela vison*	0.8	—	—	Weasel	*M. nivalis*	0.05	—	—
Ermine	*M. erminea*	0.1	—	—	Genet	*Genetta genetta*	1.6	—	VU
Long-tailed weasel	*M. frenata*	0.2	—	—			—	—	—
Least weasel	*M. nivalis*	0.05	—	—			—	—	—

Himalayas					Andes				
Brown bear	*Ursus arctos isabellinus*	95	CI	—	Spectacled bear	*Tremarctos ornatus*	125	CI	VU
Asiatic black bear	*U. thibetanus*	55	CI	CR[5]	Colpeo fox	*Pseudalopex culpaeus*	15.9	CII	—
Red panda	*Ailurus fulgens*	5.2	CI	EN	Argentine gray fox	*P. griseus*	8.3	CII	—
Tibetan wolf	*Canis lupus chanco*	36.0	CI	—	Puma	*Puma concolor*	67.5	CII	—
Wild dog	*Cuon alpinus*	11.5	CII	VU	Andean cat	*Oreailurus jacobita*	4.0	CI	VU
Golden jackal	*Canis aureus*	8.5	—	—	Geoffroy's cat	*Oncifelis geoffroyi*	3.6	CI	—
Red fox	*Vulpes vulpes montana*	3.0	—	—	Pampas cat	*O. colocolo*	3.0	—	—
Tibetan fox	*V. ferrilata*	3.5	—	—	Oncilla	*Leopardus tigrinus*	2.2	CI	LR
Leopard	*Panthera pardus*	44	CI	EN	Kodkod	*Oncifelis guigna*	2.2	—	VU
Snow leopard	*Uncia uncia*	38	CI	EN	Southern river otter	*Lontra provocax*	5.5	CI	EN
Clouded leopard	*Neofelis nebulosa*	9.4	CI	VU	Striped hog-nosed skunk	*Conepatus semistriatus*	2.4	—	—
Jungle cat	*Felis chaus*	5.0	—	—	Humboldt's skunk	*C. humboldtii*	0.3	CII[6]	—
Leopard cat	*F. bengalensis*	2.2	CI	—	Molino's skunk	*C. chinga*	1.9	—	—
Golden cat	*Catopuma temminckii*	4.2	—	LR	Lesser grison	*Galictis cuja*	1.6	—	—
Marbled cat	*Felis marmorata*	2.5	CI	DD[7]	Long-tailed weasel	*Mustela frenata*	0.2	—	—
Otter	*Lutra lutra*	6.5	CI	VU	Patagonian weasel	*Lyncodon patagonicus*	0.2	—	—
Stone marten	*Martes foina*	1.0	—	—					
Yellow-throated marten	*M. falvigula*	3.4	—	—					
Himalayan weasel	*Mustela sibirica*	1.2	—	—					
Binfurong	*Arctictis binturong*	13.0	—	—					
Spotted linsang	*Prionodon pardicolor*	4.5	—	—					
H. palm civet	*Paguma larvata*	4.1	—	—					

[1] CI: CITES Appendix I, species threatened with extinction, commercial trade in specimens is generally prohibited (UNEP-WCMC 2004).

[2] VU: IUCN Vulnerable, facing a high risk of extinction in the wild (IUCN 2001).

[3] LR: IUCN Lower Risk, does not satisfy the criteria for any of the categories Critically Endangered, Endangered, or Vulnerable (IUCN 1994)

[4] EN: IUCN Endangered, facing a very high risk of extinction in the wild (IUCN 2001).

[5] CR: IUCN Critically Endangered, considered to be facing an extremely high risk of extinction in the wild (IUCN 2001).

[6] CII: CITES Appendix II, species not necessarily now threatened with extinction but that may become so unless trade is closely controlled. Commercial trade in specimens allowed when certified not to be detrimental to the survival of the species in the wild (UNEP-WCMC 2004).

[7] DD: IUCN Data Deficient, there is inadequate information to make a direct, or indirect, assessment of its risk of extinction based on its distribution and/or population status (IUCN 2001).

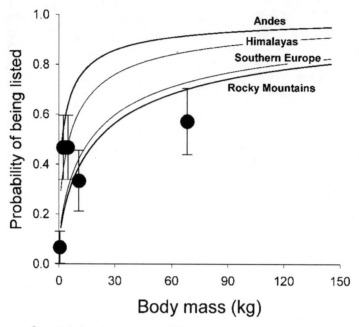

FIGURE 8.2 Relations between probability of a carnivore taxon being listed in the CITES Appendices or on the IUCN Red List and average adult female body mass, differentiating among carnivore communities in the Andes, the Himalayas, the mountains of southern Europe, and the central Rocky Mountains of North America. Filled circles and associated SE bars denote averages for quintiles and are shown to illustrate goodness of fit.

Key drivers in this model are market forces, income, culture, numbers of humans, densities of roads, and the extent of human concentration in carnivore habitat.

Prey Loss and Trapping

Figure 8.4 shows a conceptual model of relations among human and biological factors governing the endangerment of carnivores threatened primarily by trapping and prey loss. This model is relevant to small- to medium-sized carnivores that are habitat or prey specialists. Elements pertaining to rates of encounter with vehicles and humans are virtually identical to those in Figure 8.3, except that relatively high pelt values catalyze increased numbers of trappers who deliberately seek these carnivores out to kill them.

In this model, prey are usually small enough that they are not a major

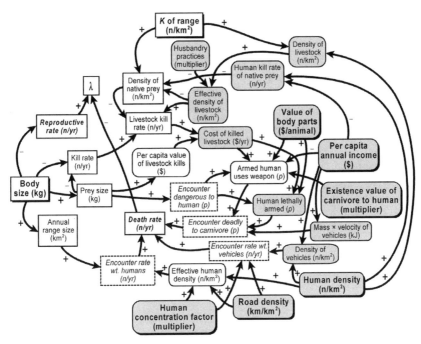

FIGURE 8.3 Conceptual model of relations among human and biological factors governing the endangerment of carnivores threatened primarily by direct human-caused deaths. Shaded boxes denote factors amenable to intervention to promote conservation.

target of human harvest for meat. Rather, prey are affected by pollution, deforestation, conversion of native vegetation to cropland, and depredation control programs. These phenomena can also cause a burgeoning of generalist predators that prey on the specialists. Human wealth and densities largely drive impacts on habitat and prey, with the effects of wealth being potentially complex.

Conclusion

Carnivore endangerment is governed by a handful of key human factors: numbers of people, densities of roads, levels of human affluence or poverty, broad-scale market forces, human culture and experiences, relative concentrations of humans, and access in favored habitats. Carnivores are almost certain to be endangered where there are high densities of roads concentrated in key habitats used by large numbers of impoverished or intolerant

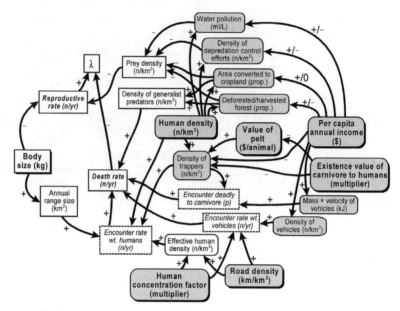

FIGURE 8.4 Conceptual model of relations among human and biological factors governing the endangerment of carnivores threatened primarily by trapping and prey loss. Shaded boxes denote factors amenable to intervention to promote conservation.

people. Such a scenario guarantees that refuges will be scarce and encounters with humans common. Moreover, encounters will likely be lethal. On the other hand, wealth breeds its own set of problems, including pollution, heightened resource demands, and large numbers of speeding vehicles on well-maintained highways.

The frequency and lethality of encounters potentially compensate for each other (Mattson et al. 1996b). Carnivore populations can potentially survive highly lethal humans given that encounters with people are rare. On the other hand, carnivores can potentially survive frequent encounters with people provided that humans are benign. However, in cases in which birth rates are low, it probably doesn't take many encounters with armed antagonistic people to drive a population of carnivores to extinction. The importance of human lethality in governing the fates of especially large carnivores is evinced by the rapid and widespread extirpations of wolves and grizzly bears in the contiguous United States at a time when there were few roads and relatively few people. This suggests a primary role for human values, perspectives, and behaviors in carnivore conservation. Even so, there are obvious limits to the numbers of tolerant and otherwise accepting people who can concentrate in important carnivore habitat without having a serious negative impact on predators and their prey (Mattson et al. 1996a).

Primary drivers aside, there are a number of salutary measures that can contribute to the conservation of carnivores. Livestock producers can be compensated for losses so that they have less incentive to kill depredating predators. Harvest of native prey can be curbed so that predators will be less motivated to kill livestock. Livestock numbers can be reduced. Producers can convert to rearing larger-bodied or more vagile animals. Husbandry practices can be instituted that make small-bodied livestock less vulnerable. Programs that control or eliminate depredating herbivores can be curbed or eliminated so that prey populations are conserved. Sanctions can be imposed to influence the value of body parts on national and international markets. Measures can be enacted and resources allocated for enforcement to protect carnivores outright. Natural vegetation and disturbance regimes can be restored and protected. Free-flowing water can be cleaned. Roads can be fenced and speed limits imposed. As in many national parks, humans can be disarmed. That said, humans can become so impoverished, so numerous, or so intolerant that no amount of finagling can save carnivore populations. As much as attention to secondary factors can help, primary drivers are inescapable.

Human factors aside, there are certain biological traits that influence risk of endangerment among carnivores. Species averaging larger than 8 kg as adults or specializing on a limited number of prey in a limited number of habitats are most vulnerable. By being attuned to these biological traits, managers can better assess the vulnerabilities of different species in novel situations and better understand the nature of associated conservation problems. In limited circumstances, these biological traits can be manipulated to promote conservation, as in instances in which fences are installed or availability of prey or carrion is augmented to curb movements.

Whatever the human and biological factors, the identification of threats and the articulation of potentially efficacious tactics and strategies is but one step in carnivore conservation. The promotion, adoption, implementation, and appraisal of governing policies necessarily follow (Clark 2002). Crafting an effective conservation policy process requires not only knowledge about humans and human social systems but also considerable skill in operating in policy arenas. Without such skill and without such knowledge, no amount of concern or knowledge about the carnivores themselves will be sufficient to save them.

Literature Cited

Bashares, J. S., P. Arcese, and M. K. Sam. 2001. "Human demography and reserve size predict wildlife extinction in West Africa," *Proceedings of the Royal Society of London* B 268:2473-2478.

Brown, D. E. 1983. *The Wolf in the Southwest: The Making of an Endangered Species.* University of Arizona Press, Tucson.

———. 1996. *The Grizzly in the Southwest: Documentary of an Extinction.* University of Oklahoma Press, Norman.

Carroll, C., R. F. Noss, and P. C. Paquet. 2001. "Carnivores as focal species for conservation planning in the Rocky Mountain region," *Ecological Applications* 11:961-980.

Ceballos, G., and J. H. Brown. 1995. "Global patterns of mammalian diversity, endemism, and endangerment," *Conservation Biology* 9:559-568.

Clark, T. W. 2002. *The Policy Process: A Practical Guide for Natural Resource Professionals.* Yale University Press, New Haven.

Clevenger, A. P., B. Chruszcz, and K. E. Gunson. 2001. "Highway mitigation fencing reduces wildlife-vehicle collisions," *Wildlife Society Bulletin* 29:646-653.

Crabtree, R. L., and J. W. Sheldon. 1999. "Coyotes and canid coexistence in Yellowstone," in *Carnivores in Ecosystems: The Yellowstone Experience,* ed. T. W. Clark, A. P. Curlee, S. C. Minta, and P. M. Kareiva, 127-163. Yale University Press, New Haven.

Czech, B., P. R. Krausman, and P. K. Devers. 2000. "Economic associations among causes of species endangerment in the United States," *BioScience* 50:593-601.

Dudley, J. P., J. R. Ginsberg, A. J. Plumptre, J. A. Hart, and L. C. Campos. 2002. "Effects of war and civil strife on wildlife and wildlife habitats," *Conservation Biology* 16:319-329.

Ferreras, P., J. J. Aldama, J. F. Beltran, and M. Delibes. 1992. "Rates and causes of mortality in a fragmented population of Iberian lynx (*Felis pardina*) Temminck, 1824," *Biological Conservation* 61:197-202.

Forman, R. T. T., and L. E. Alexander. 1998. "Roads and their major ecological effects," *Annual Review of Ecology and Systematics* 29:207-231.

Fox, J. L. 1991. "Status of the snow leopard *Panthera uncia* in northwest India," *Biological Conservation* 55:283-298.

Funnell, D., and R. Parish. 2001. *Mountain Environments and Communities.* Routledge, New York.

Gittleman, J. L. 1985. "Carnivore body size: Ecological and taxonomic correlates," *Oecologia* 67:540-554.

Gittleman, J. L., and P. H. Harvey. 1982. "Carnivore home-range size, metabolic needs and ecology," *Behavioral Ecology and Sociobiology* 10:57-63.

Gottelli, D., and C. Sillero-Zubiri. 1992. "The Ethiopian wolf: An endangered endemic canid," *Oryx* 26:205-214.

Griffiths, H. I. 2000. *Mustelids in a Modern World: Management and Conservation Aspects of Small Carnivore and Human Interactions.* Backhuys, Leiden, Netherlands.

Hall, B., and M. L. Kerr. 1991. *1991-1992 Green Index: A State-by-State Guide to the Nation's Environmental Health.* Island Press, Washington, DC.

Hargis, C. D., J. A. Bissonette, and D. L. Turner. 1999. "The influence of forest fragmentation and landscape pattern on American martens," *Journal of Applied Ecology* 36:157–172.

Hennemann, W. W. III. 1983. "Relationship among body mass, metabolic rate and the intrinsic rate of natural increase in mammals," *Oecologia* 56:104–108.

Houghton, R. A. 1994. "The worldwide extent of land-use change," *BioScience* 44:305–313.

IUCN. 1994. IUCN Red List Categories and Criteria (version 2.3). www.redlist.org/info/categories_criteria1994.html#categories.

IUCN. 2001. IUCN Red List Categories and Criteria (version 3.1). www.iucnredlist.org/info/categories_criteria2001.html#categories.

IUCN. 2003. *2003 IUCN Red List of Threatened Species*. www.redlist.org.

IUCN Canid Specialist Group. Species accounts. www.canids.org/SPPACCTS/sppaccts.htm.

IUCN Cat Specialist Group. Species accounts. http://lynx.uio.no/catfolk.

Jackson, R. M., G. G. Ahlborn, M. Gurung, and S. Ale. 1996. "Reducing livestock depredation losses in the Nepalese Himalaya," in *Proceedings of the 17th Vertebrate Pest Conference* ed. R. M. Timm and A. C. Crabb, 241–247. University of California, Davis.

Kaczensky, P., F. Knauer, M. Jonozović, T. Huber, and M. Adamič. 1996. "The Ljubljana-Postojna highway: A deadly barrier for brown bears in Slovenia," *Journal of Wildlife Research* 1:263–267.

Kellert, S. R. 1985. "Public perceptions of predators, particularly the wolf and coyote," *Biological Conservation* 31:167–189.

Kerr, J. T., and D. J. Currie. 1995. "Effects of human activity on global extinction risk," *Conservation Biology* 9:1528–1538.

Landa, A., K. Gundrangen, J. E. Swenson, and E. Røskaft. 1999. "Factors associated with wolverine *Gulo gulo* predation on domestic sheep," *Journal of Applied Ecology* 36:963–973.

Logan, K. A., and L. L. Sweanor. 2000. "Puma," in *Ecology and Management of Large Mammals of North America*, ed. S. Demarais and P. R. Krausman, 347–377. Prentice Hall, Upper Saddle River, NJ.

Mason, C. F., and S. M. Macdonald. 1986. *Otters: Ecology and Conservation*. Cambridge University Press, Cambridge, U.K.

Massolo, A., and A. Meriggi. 1998. "Factors affecting habitat occupancy by wolves in northern Appenines (northern Italy): A model of habitat suitability," *Ecography* 21:97–107.

Mattson, D. J. 1990. "Human impacts on bear habitat use," *International Conference of Bear Research and Management* 8:33–56.

Mattson, D. J., S. Herrero, R. G. Wright, and C. M. Pease. 1996a. "Designing and managing protected areas for grizzly bears: How much is enough?" in *National Parks and Protected Areas: Their Role in Environmental Protection*, ed. R. G. Wright, 133–164. Blackwell Science, Cambridge, MA.

———. 1996b. "Science and management of Rocky Mountain grizzly bears," *Conservation Biology* 10:1013–1025.

Mattson, D. J., and T. Merrill. 2002. "Extirpations of grizzly bears in the contiguous United States, 1950–2000," *Conservation Biology* 16:1123–1136.

Mattson, D. J., T. Merrill, and L. Craighead. In review. "Predicting umbrella effects: A multidimensional method applied to carnivores in Montana and Idaho, USA." *Conservation Biology*.

Medina, A. F. A Guide to Andean Countries. www.ddg.com/LIS/aurelia/titpag.htm.

Miller, B., R. P. Reading, and S. Forrest. 1996. *Prairie Night: Black-Footed Ferrets and the Recovery of Endangered Species*. Smithsonian Press, Washington, DC.

Miquelle, D. G., E. N. Smirnov, T. W. Merrill, A. E. Myslenkov, H. B. Quigley, M. G. Hornocker, and B. Schleyer. 1999. "Hierarchical spatial analysis of Amur tiger relationships to habitat and prey," in *Riding the Tiger,* ed. J. Seidensticker, S. Christie, and P. Jackson, 71–99. Cambridge University Press, New York.

Murray, D. L., C. A. Kapke, J. F. Everman, and T. K. Fuller. 1999. "Infectious disease and the conservation of free-ranging large carnivores," *Animal Conservation* 2:241–254.

Oli, M. K., I. R. Taylor, and M. E. Rogers. 1994. "Snow leopard *Panthera uncia* predation of livestock: An assessment of local perceptions in the Annapurna Conservation Area, Nepal," *Biological Conservation* 68:63–68.

Pease, C. M., and D. J. Mattson. 1999. "Demography of the Yellowstone grizzly bears," *Ecology* 80:957–975.

Rao, M., and P. J. K. Gowan. 2002. "Wild-meat use, food security, livelihoods, and conservation," *Conservation Biology* 16:580–583.

Reading, R. P., H. Mix, B. Lhagvasuren, and N. Tseveenmyadag. 1998. "The commercial harvest of wildlife in Dornod Aimag, Mongolia," *Journal of Wildlife Management* 62:59–71.

Reig, S., L. Cuesta, F. Palacios, and F. Barcena. 1992. "Status of the wolf in Spain," in *Global Trends in Wildlife Management.* vol. 2, ed. B. Bokek, K. Perzariowski, and W. L. Regelin, 371–375. Swiat Press, Krakow, Poland.

Roelke-Parker, M. E., L. Munson, C. Packer, R. Kock, S. Cleaveland, M. Carpenter, S. J. O'Brien, A. Pospischil, R. Hofmann-Lehmann, H. Lutz, G. L. M. Mwamengele, M. N. Mgasa, G. A. Machange, B. A. Summers, and M. J. G. Appel. 1996. "A canine distemper virus epidemic in Serengeti lions (*Panthera leo*)," *Nature* 379:441–445.

Ruggiero, L. F., K. B. Aubry, S. W. Buskirk, G. M. Koehler, C. J. Krebs, K. S. McKelvey, and J. R. Squires. 1999. *Ecology and Conservation of Lynx in the United States.* U.S. Forest Service General Technical Report RMRS-GTR-30WWW. University of Colorado Press, Boulder.

Schaller, G. B. 1998. *Wildlife of the Tibetan Steppe.* University of Chicago Press, Chicago.

Seidensticker, J. 1997. "Saving the tiger," *Wildlife Society Bulletin* 25:6–17.

Stahl, P., J. M. Vandel, V. Herrenschmidt, and P. Migot. 2001. "The effect of removing lynx in reducing attacks on sheep in the French Jura Mountains," *Biological Conservation* 101:15–22.

Terborgh, J., J. A. Estes, P. Paquet, K. Ralls, D. Boyd-Heger, B. J. Miller, and R. F. Noss. 1999. "The role of top carnivores in regulating terrestrial ecosystems," in *Continental Conservation: Scientific Foundations of Regional Reserve Networks,* ed. M. E. Soulé and J. Terborgh, 39–64. Island Press, Washington, DC.

Trombulak, S. C., and C. A. Frissell. 2000. "Review of ecological effects of roads on terrestrial and aquatic communities," *Conservation Biology* 14:18–30.

UNEP-WCMC. 2004. *UNEP-WCMC Species Database: CITES-Listed Species.* http://sea.unep-cmc.org/isdb/CITES/Taxonomy/fa_user.cfm.

Weaver, J. L., P. C. Paquet, and L. F. Ruggiero. 1996. "Resilience and conservation of large carnivores in the Rocky Mountains," *Conservation Biology* 10:964–976.

Woodroffe, R. 2001. "Strategies for carnivore conservation: Lessons from contemporary extinctions," in *Carnivore Conservation,* ed. J. L. Gittleman, S. M. Funk, D. W. Macdonald, and R. K. Wayne, 61–92. Cambridge University Press, Cambridge, U.K.

Woodroffe, R., and J. R. Ginsberg. 1998. "Edge effects and the extinction of populations inside protected areas," *Science* 280:2126–2128.

Wootton, J. T. 1987. "The effects of body mass, phylogeny, habitat, and trophic level on mammalian age at first reproduction," *Evolution* 41:732–749.

PART 3

Coexistence in Political Landscapes

Parts 1 and 2 have presented numerous examples of real conflict between people and predators in rural and developed areas, and additional cases of potential and perceived conflict. It should also be apparent that most of the authors feel there are workable solutions, through technology and innovative management strategies, to the concrete problems of living with carnivores. But the resolution of these conflicts is made more daunting by the complex political realities encountered when managing carnivores.

Because of their often large size, and indeed their very nature as carnivores—predators that can do economic damage to agricultural interests, compete with hunting interests for wild game, or even threaten, or appear to threaten, human safety—the order Carnivora provokes feelings rarely seen with other taxa. Compounding the challenging on-the-ground realities of managing carnivores are the positive and negative emotions elicited by such species as the wolf, a creature shrouded in myth and superstition. These real, human emotions can cloud the political decisions surrounding already complicated management decisions. Yet with irrefutable scientific proof that carnivores play a key role in balancing our ecosystems, emotions cannot dominate the discussion of carnivore management in North America. In the four chapters of Part 3, authors who face these issues in their professional lives discuss the need to find that elusive balance between addressing human concerns and ensuring that the biological needs of carnivores are met.

Chapter 9, "Dispersal and Colonization in the Florida Panther: Overcoming Landscape Barriers—Biological and Political," by David S. Maehr, explores the heated topic of Florida panther recovery in south Florida. Maehr asserts that although there are biological, or landscape, barriers to expansion of the perilous Florida panther population, the political roadblocks are the primary impediment to achieving successful and long-term cat conservation in the state.

Chapter 10, "State Wildlife Governance and Carnivore Conservation," by Martin Nie, continues the examination of the state role in carnivore management and how various sociopolitical interests have long governed the discourse on this issue. Interestingly, management paradigms change over the years as public sentiment toward predators changes and various and diverse constituencies gain political power. Nie questions whether this is an appropriate approach to management of such an important natural resource.

Chapter 11, "Conserving Mountain Lions in a Changing Landscape," by Christopher M. Papouchis, examines the role of various state wildlife agencies in mountain lion conservation and management over time and suggests what can be done to protect this species into the future.

Chapter 12, "Restoring the Gray Wolf to the Southern Rocky Mountains: The Anatomy of a Campaign to Resolve a Conservation Issue," by Michael K. Phillips, Rob Edward, and Tina Arapkiles, demonstrates how a complex, holistic approach is necessary in restoring a large, controversial predator. Though the wolf as a species may always have its detractors, those proposing its return to the lower forty-eight states can help minimize public anger by ensuring that all those vested in the process have a voice in the process.

Thus it should become clear that political considerations have as much, and sometimes even more, impact on conservation initiatives and long-term management of North America's large predators as do biological implications. In a world with a growing human population that places increasing pressure on our natural world, the key to successful cohabitation with and conservation of predators will clearly entail close examination of our emotional and political motivators. By honestly assessing these issues, and addressing them in tandem with management realities, we will have better success at moving from conflict to conservation.

CHAPTER 9

Dispersal and Colonization in the Florida Panther: Overcoming Landscape Barriers—Biological and Political

David S. Maehr

Dispersal is of such overriding importance to the conservation of the Florida panther (*Puma concolor coryi*) that virtually all of its demographic and genetic ills (real or imagined) could be repaired by restoring connectivity with other populations. Certainly, a cynic might observe that the lack of another nearby population reduces such an argument to an optimistic daydream, especially if one is willing to accept the status quo, but the power of dispersal in maintaining populations has long been recognized (Lidicker and Caldwell 1982). Regardless, successful recovery will occur only when the panther is demographically secure—an outcome that must include dispersal patterns that are representative of the species. Elsewhere, Maehr et al. (2002) and Maehr and Caddick (1995) have detailed the spatial, temporal, and density attributes in North America's only southeastern population of the species and concluded that the panther was behaviorally capable of colonizing vacant habitat and that reproduction in core range was sufficient to support population growth. Why then, has the population failed to expand significantly on its own even after the introduction of Texas cougars into south Florida?

Only artificial landscape barriers seem to be implicated in preventing the panther from colonizing large areas of Florida. Perhaps more serious obstacles are a geocentric view of recovery by some land managers, and the resistance of agencies and a vocal minority of the public to the idea of repatriating a wide-ranging carnivore to portions of its historic range. This chapter examines the ecological and political realities of Florida panther dispersal in a development state, and offers suggestions for harnessing this aspect of panther ecology to enable expansion in this small, isolated population.

Background

For more than a century, the Florida panther has been spatially constrained by public works projects that began with the dredging of a navigable waterway between Lake Okeechobee and the Gulf of Mexico at the end of the nineteenth century (Maehr 1997a). Subsequent agricultural and urban development exacerbated the effects of a peninsular distribution already restricted by coastlines, the largest freshwater marsh in the southeast, and a 1,770 km^2 lake. The culmination of subsequent modifications is the current pattern of landscape islands, denatured and isolated by humanity's best efforts to convert a harsh, wet, and subtropical landscape into tomato farms, cattle ranches, resorts, and recreation areas.

By the time the remnants of a population that once spanned the entire southeastern coastal plain were discovered in extreme south Florida, the panther had contracted its range to just a handful of counties. Although the coastal areas of this region were already heavily urbanized when recovery efforts began in the early 1970s, the interior still supported a vast network of wilderness associated with some of the East's largest nature preserves. Certainly, the wilderness that would become Everglades National Park and Big Cypress National Preserve were important to buffering the effects of a rapidly growing human population (Figure 9.1), but the as yet undeveloped private lands were even more important. These lands, however, were under increasing pressure to support human residences, an expanding citrus industry, and winter vegetable production (Maehr 1990). This complex of private property is part of the most productive portion of the panther's modern range (Maehr 1997a).

With the former east-west forest connection obliterated by the conversion of Lake Okeechobee's south shore delta to sugar cane (McCally 1999), development in southwest Florida seemed poised to constrict the panther into an impossibly small population fragment. Despite these pressures, two decades of fieldwork documented a resilient population that exhibits remarkably typical demographics and behavior (Maehr and Caddick 1995; Maehr 1997c). As has been noted elsewhere, "the only notable demographic differences between the Florida panther and most other well-studied cougar populations" are "an earlier age of independence and a high degree of circular dispersals" (Maehr et al. 2002). Although dispersal

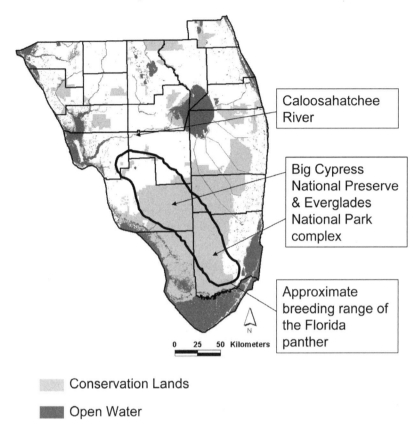

FIGURE 9.1 The southern peninsula of Florida includes the traditionally occupied range of the Florida panther south of the Caloosahatchee River, where the landscape is dominated by public lands, and a potential colonization zone north of the river. The potential colonization zone is dominated by private lands and an arrangement of forest cover that differs from that of known occupied range. This means that its potential to support a panther population is unproven. The eastern third of the Caloosahatchee River is an artificial waterway, dredged in the late 1800s. It and the rest of the channel may represent a barrier to female panther movements.

can be easily disrupted by a variety of management activities and land uses, it should also be easily facilitated by landscape repair (MacDonald and Johnson 2001).

A Conceptual Framework

Whereas the phenomenon of dispersal is the subject of a number of theories that examine its phenology and evolutionary history (Lidicker and

Caldwell 1982; Swingland and Greenwood 1983; Clobert et al. 2001), an understanding of the conditions under which it occurs is of more importance to this discussion. Panther dispersal, as in most solitary carnivores, is fundamentally different from that found in many plants and invertebrates, which may flood local areas with millions of propagules that are dependent on wind or water currents for transport to potential colonization sites. These periodic pulses of reproductive potential are often unrelated to the density of the parent population, may contain propagules of both sexes (or self-fertilizing individuals), and maintain a constant potential for colonization of new ranges. Successful establishment for such species is heavily dependent on finding a suitable spot on which to "land."

Florida panther dispersal, on the other hand, appears to be directly related to the density of the source population, and is exhibited primarily by males. Further, reproductively successful dispersal depends on the presence of females at the termination of a dispersal attempt (Maehr et al. 2002). Sadly, for an isolated population such as the Florida panther, its range can expand only if males and females colonize vacant range together, an event that has not occurred in over three decades of monitoring this endangered subspecies.

Pressure, Resistance, Sex, and Motivation

Howard (1960) and Ims and Hjermann (2001) suggested that dispersal was a product of either innate or environmental influences. The pattern of panther dispersal is likely a combination of both, and is certainly affected by anthropogenic changes to the landscape. In the absence of severe landscape denaturement it may be helpful to think of the panther dispersal phenomenon as an interacting set of physical laws and behavior.

For males, the rate and direction of movement away from home depends on pressure and resistance. Much like Boyle's law, which states the volume of an ideal gas is inversely proportional to pressure, panther population density (volume) will be relieved at high densities by the exodus of dispersing individuals. Accordingly, the pressure behind this escape may be represented by intolerant resident adult males and subsequent negative encounters. The motivation for an animal to move away from its natal territory is likely a combination of an inherent tendency for young males to wander and the influence of more dominant, intolerant individuals that are more strongly motivated to defend mating opportunities than the newcomer is to wrest them away.

The distance and direction moved may also be related to pressure but is further influenced by the resistance of the landscape (Wiens 2001). Ohm's law states that current is inversely proportional to resistance (landscape features) and proportional to the applied electromotive force (motivation). From the panther's perspective, resistance to movement appears to increase as forest cover decreases (Meegan and Maehr 2002)—that is, fragmented landscapes with little forest cover appear to handicap panther movements.

A more organic comparison may be Laplace's law, whereby "the energy expended in ejecting a given quantity of blood from the heart depends on the efficiency of contraction, the pressures developed, and the size and shape of the heart" (Randall et al. 2002). The usefulness of this comparison is in the incorporation of both pressure and structural elements in determining how a fluid moves through the circulatory system. This analogy was demonstrated on the panther landscape by recent dispersal events whereby at least three subadult males negotiated a barrier-rich landscape before crossing the Caloosahatchee River (Maehr et al. 2002), an artificial waterway that until 1998 had appeared to function as a movement barrier (Maehr 1997a). Previous to this, all male dispersal events were circular and most were interrupted by mortality before they were complete (Maehr et al. 1991a; Maehr 1997a), most likely because population density had not reached the threshold required to drive young males across a denatured landscape (Maehr et al. 2002).

Regardless of the pressure behind panther dispersal, immigrant males become "a rather tragic procession of refugees, with all the obsessed behavior of the unwanted stranger in a populous land, going blindly on to various deaths" (C. Elton, as quoted in Gray 1948). Indeed, successful male panther recruitment happens only to the extent that vacancies arise. Those individuals that time their independence in the absence of a vacant home range are destined to wander through marginal habitat that is unoccupied by conspecifics of the opposite gender, or they must battle standing residents for breeding rights—usually with fatal consequences to the newcomer (Maehr et al. 1991a).

Although the Florida panther population had already been exhibiting a gradual increase (Figure 9.2), including marginal range expansion through the early 1990s (Maehr 1997a), the introduction of female cougars from Texas in 1995 promoted additional reproduction and increased density in occupied range that was likely already at carrying capacity (Maehr et al. 2002). This became the pressure and increased motivation behind the dis-

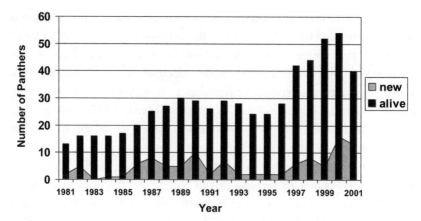

FIGURE 9.2 Estimates of the number of known Florida panthers from 1981 through 2001 are a function of population size and capture effort. The estimated ages at capture were used to determine when each animal entered the population. Although these figures are not without bias, they suggest a gradual increase in population size that began in the mid-1980s, before the introduction of Texas cougars in 1995. A reduction in the rate of increase between 1991 and 1995 was due primarily to a shift in focus on establishing a captive breeding program and the removal of 10 kittens, followed by personnel constraints and fewer capture attempts. The steep increase after 1996 is related to the focus by FFW-CC on capturing and monitoring the offspring of introduced Texas females. In addition, capture efforts through 1994 were restricted to approximately 3 months per year, whereas in subsequent years capture activities were conducted throughout the year.

persers that crossed a capillary-like network of remnant forest and facilitated the spanning of the heretofore barrier of the Caloosahatchee River.

The dispersal events themselves displayed remarkably similar trajectories (Meegan and Maehr 2002) despite varying pedigrees of the cats (the first was made by a non-Texas hybrid male), and separations in time of about one year between each of the events (Maehr et al. 2002) (Figure 9.3). It is likely that since the initiation of radiotelemetry studies in the 1980s, the effective resistance of the landscape has increased as new roads, more traffic, and concomitant development have rendered an already patchy landscape more dangerous and impenetrable (Figure 9.4). On the other hand, although the south-central Florida landscape supports different patterns of forest cover, it also has highways with lower traffic volumes than many that are in the heart of occupied range. Thus, once home ranges are established in this potential zone of colonization, resident panthers may exhibit lower levels of highway mortality and less severe habitat qual-

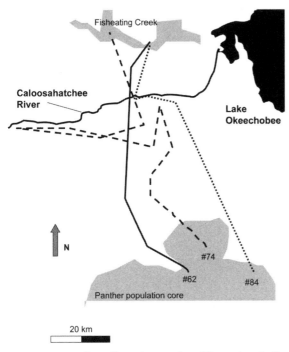

FIGURE 9.3 Three dispersing panthers (ID numbers indicated at their approximate dispersal origins) traveled north from the core population in southwest Florida to the fisheating Creek area in south-central Florida between 1998 and 2000 (adapted from Meegan and Maehr 2002). Each of these radio-collared males crossed the >100 km long Caloosahatchee River within a 2 km span suggesting that this consistent (non-random) selection was influenced by landscape features such as forest cover and human habitation (i.e., selection for and against, respectively).

ity reduction as the result of edge effects associated with roads (Orlando 2003).

Regardless, increased pressure in the source population has forced sweepstakes-like movements across a denatured landscape instead of the land-bridge type of migration that permits all species to disperse, to travel in multiple directions, and to maintain balanced populations (Simpson 1940; Harris and Scheck 1991); such migrations were typical of Florida panther dispersal before humans dramatically altered the landscape. Had another population existed between Labelle and Orlando at the time of recent male efforts to colonize new range, successful dispersal would likely have occurred. Instead, of the 4 animals that were radio tracked north of the Caloosahatchee River, 3 died of various causes, whereas the fourth

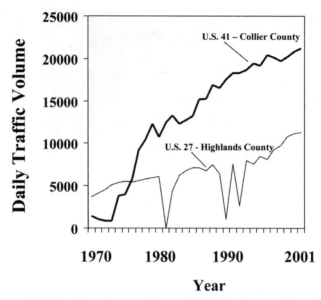

FIGURE 9.4 Traffic volume increases in occupied (Collier County—heavy line) and potential (Highlands County—narrow line) panther range in southwest and south-central Florida, respectively. Values represent the mean daily traffic volumes for 3 highway monitoring locations in potential or occupied panther range from 1970 through 2001. Source: Data derived from annual Florida Department of Transportation traffic count surveys conducted for a single day at each location.

negotiated a >500 km circle before his radio failed. None of them was female, and therein lurks the problem for population colonization and expansion in the Florida panther.

Female panther dispersal is philopatric—that is, young females tend to establish home ranges that overlap with those of their mothers regardless of source population density. At least one Florida panther took this pattern to an extreme by raising her first litter entirely within the boundaries of her natal range (Maehr 1997a). The average effective dispersal distance among females is just over 11 km, and only one has been documented to exceed 30 km (Maehr et al. 2002). Even under the increased pressure of an artificially high population density (Maehr and Lacy 2002), female dispersal distances, unlike those of males, have not changed since the introduction of Texas cougars (Maehr et al. 2002). Although longer female dispersals occur (Logan and Sweanor 2001), the rarity of such events likely precludes coincident colonization with a male north of the Caloosahatchee River.

The relative plasticity of male panther dispersal suggests that the species evolved within a system of well-connected populations, and that even the relative isolation of the Florida peninsula was insufficient to prevent risk-taking males from maintaining gene flow to reduce inbreeding that is naturally exacerbated by philopatry.

The challenge facing panther recovery is in restoring a population of sufficient size for dispersal to promote connectivity in a metapopulation that is large enough to reduce the need for frequent intervention to address genetic concerns (Hansson 1991). At some point, females will need to become a part of the colonization process before the vision of recovery can even begin to unfold. Until this happens, Florida panther dispersal will remain frustrated and ineffective (Maehr et al. 2002).

The Obstacles of Policy and Agency Will

Unfortunately, although opportunities exist for landscape solutions to panther recovery problems, their potential for application is directly related to the collective will of individual people and the organizations they represent. Elsewhere I have referred to "get real"-ism, the widespread malaise that infects natural resource agencies around the world and stifles grass-roots efforts to correct environmental mistakes (Maehr 2001). It is a philosophy that accepts nothing better than the status quo and denies the benefits of reconnecting fragmented populations of large terrestrial vertebrates such as the Florida black bear (*Ursus americanus floridanus*), or in expanding the single breeding population of *Puma concolor* in the East. It is a problem that faces Florida panther recovery today. For the panther, success or failure is determined by the people, especially those in decision making and public-contact positions, who are charged with the task of recovery. In this context, individuals involved at high levels in government can, themselves, become limiting factors in the survival of species (Maehr 1997b).

As I have written earlier (Maehr 2001), carnivores are generally recognized as the most difficult species to restore because of the difficulties in reversing the human social landscape that caused extinction in the first place (Griffith et al. 1989; Stanley Price 1989; Yalden 1993). The cultural antipathy toward large, free-ranging, meat-eating animals is very apparent when there is the perception that they will hide behind houses and steal away children. This sort of resistance is not unusual. It has happened with wolves in Michigan (Yalden 1993), with panthers in Florida (Belden and

Hagedorn 1993), and with elk in Kentucky (Maehr et al. 1999). In these situations, vocal minorities opposed reintroduction and succeeded in constraining restoration efforts, especially for the two carnivores. But restoration attempts have proceeded despite pockets of negativism. Gray wolves in the American West and an exhaustive effort to return the grizzly bear to a portion of the northern Rocky Mountains demonstrate that the proper landscape and the dogged efforts of individuals are keys to opening the crate and releasing a flash of fur into the forest.

But the establishment of a breeding panther population north of the Caloosahatchee River can hardly be called reintroduction—it is much less than that from both policy and demographic perspectives. Panthers are already there—it is occupied range, but without documented reproduction. This fact may become a technical stumbling block for those who would rather deny the possibilities than seize the opportunities in recovering the Florida panther.

Translocation has been urged for a variety of species as a practical and simple solution when the potential for successful expansion through dispersal is unlikely or impossible (Hanski 1999; MacDonald and Johnson 2001). In addition, the precedent in south Florida has already been established by moving a female into an area of desired population expansion (E. D. Land, Florida Fish and Wildlife Conservation Commission, pers. comm.). A simple first step in testing the ability of the south-central Florida landscape to support a breeding panther population would be to relocate females from the south Florida population. This population has already exhibited the ability to sustain removals for the establishment of the now-defunct captive breeding program. Further, the ongoing long-distance dispersals of males, and a recent increase in female deaths caused by males (E. D. Land, Florida Fish and Wildlife Conservation Commission, pers. comm.) suggest that south Florida is beyond carrying capacity (Maehr et al. 2002). Translocation of a female would simply be a within-range management adjustment to overcome the artificial landscape barrier of a dredged waterway and the loss of forest caused by agriculture and other human activities. This is an action that managers should undertake even while a new recovery plan is being composed. To delay could mean missing the window of opportunity that benign south Florida environmental conditions and introduction-induced kitten production have already created (Maehr and Lacy 2002).

While it is commendable that the U.S. Fish and Wildlife Service has convened a new Florida panther recovery team, its makeup of more than 30 members promises to be an exercise in satisfying a diversity of political positions before progress is made in actual recovery. Political wrangling has been a problem even for smaller groups charged with recovery during the previous two decades (Alvarez 1993). In particular, limited leadership by the Florida Fish and Wildlife Conservation Commission (FFWCC) in efforts to conserve panther habitat on private lands (Alvarez 1993) will further alienate landowners. These landowners must increasingly contend with inflexible development regulations as well as permitting authorities who are constrained by shrinking budgets, poor morale, and little knowledge of carnivore biology or landscape ecology. There seems to be more interest in forcing landowners to accept panthers on private lands than in enlisting cooperation through a variety of approaches that have been recommended by many authors (Rusmore et al. 1982; Adams and Dove 1989; Noss et al. 1997; Meegan and Maehr 2002).

Because there are likely insufficient funds available for the government to buy all the land necessary to recover and expand panther range, the cooperation of landowners is essential (Maehr 1990). Although visionary planning for conservation lands in Florida would set aside nearly 60% of the state in some form of protection, Hoctor et al. (2000) recommended that conservation easements become a more utilized tool for doing so. This suggests a future in which biodiversity conservation is heavily dependent on private land.

The lack of leadership is not restricted to state agencies. For example, a geocentric (i.e., "our land is the center of all that is important") view among many in the National Park Service endangers conservation efforts where they are truly needed—at the fringes of public lands and the urban-wilderness interface. A move away from panther monitoring activities that appear more directed at the personal satisfaction of individual staff is a needed step in the evolution of Florida panther recovery. It is an amazing feat that the Big Cypress National Preserve has been in the panther research business for nearly two decades without a single published summary of the monitoring that has occurred there since the 1980s.

Divisiveness and the lack of a shared vision among the state and federal agencies involved in panther research, management, and administration further handicaps a coordinated approach to recovery. Recently, the tradition of a single capture team for all panthers was scuttled in favor of a multiagency approach using multiple capture crews, despite the lack of compelling reasons for doing so. Ken Alvarez, a former member of the Florida Panther Recovery Team and the governor-appointed Florida Panther Technical Advisory Council, examined the dynamics of interagency panther recovery and offered a dismal outlook in his 1993 book. With respect to the development of the recovery plan (USFWS 1987), he observed:

> The recovery plan exercise allowed each faction to eliminate or reword anything it found disagreeable—prior to having the draft go out for public review. It could be argued that legitimate editorial functions were involved, but the administrators had not made a disciplined analysis of panther survival. They were each trying to fit the design to long-standing notions of agency self-interest. Five years after its beginning, the effort to recover the panther in southern Florida would still be in motion, churning out optimistic statements and spending huge sums of money, with no idea where it was going. . . . The Fish and Wildlife Service would finally resort to contracting with a group of independent technical experts from around the country to try and impose sanity on the course of affairs. (Alvarez 1993)

Alvarez concluded his view of the new recovery plan by observing that

> the land management agencies will not come to grips with the core issues and instead paste them over with inane news releases to assuage public opinion. Innovative and meaningful ideas are ignored, avoided or squelched. Those who maneuver to uphold the status quo are given awards. Words and forms triumph over substance. The Florida panther recovery program has been a masterpiece of bureaucratic art. (Alvarez 1993)

Now, 10 years later, after a half-dozen panthers have exhibited the ability to cross once-believed landscape barriers into once-occupied range, denial still reigns. In response to the recent death of a panther in the vicinity of Tampa, Florida, a FFWCC spokesperson chided suggestions from outsiders (i.e., nonagency scientists) that females be moved north of the river by saying, "These biologists are not living in the real world. . . . Panther management is not just a biological issue. It's burdened with huge political and legal issues that require significant planning and large land purchases to ensure that efforts succeed" (Alberto 2003). No one disputes the existence of these sociopolitical challenges, but how many more decades are needed to move panther recovery and management beyond the Caloosahatchee River? And how long will it take for managers to realize that recovery will not happen without vast areas of privately owned, interconnected panther habitat? Enough land purchases to acquire all the habitat needed to protect the panther are not forthcoming, even with the state of Florida's aggressive land conservation program, Florida Forever.

While panthers continue to promote their own recovery through long-distance but frustrated dispersal, the agencies have appeared to reduce the likelihood of success by eliminating valuable options. Why, after the investment of several years studying an experimental cougar population in north Florida, and removing 10 kittens from the wilds of south Florida, did state wildlife authorities withdraw completely from the panther reintroduction experiment? If the will had truly existed to reestablish additional populations of panther in Florida, why was there no aggressive education program initiated to create a future social atmosphere that was more supportive of reintroduction? And why today, as young male panthers hurl themselves into the outer range limits of south-central Florida, have not these same agencies seized upon this biological momentum to provide added security to a small and profoundly isolated population? The landscape itself was once viewed as the primary barrier to successful dispersal and population expansion (Maehr et al. 1991b). More than a decade later, the larger threat is the intransigence and inaction of agencies and their representatives.

How to Obtain Recovery

Shrader-Frechette and McCoy (1993) observed that developers might have a reasonable argument for developing panther habitat "(1) if there were a better use of conservation dollars, all things considered, than to spend them on a "flagship" taxon . . . such as the panther; (2) if there were no more available land to be developed; (3) if the current land had been used as efficiently as possible; and (4) if current use of the panther habitat were necessary to avert some great harm such as human starvation." None of these conditions appears to pertain in either south Florida or south-central Florida, so panther conservation provides a greater contribution to the common good than does development (Shrader-Frechette and McCoy 1993).

Unfortunately, development in south Florida will continue indefinitely to some degree in occupied or otherwise good panther habitat. Thus, a grave responsibility falls on the shoulders of regulatory authorities. Although quantitative methods for evaluating habitat and mitigating panther habitat loss have been suggested (Maehr and Deason 2002), such approaches do little to take into consideration the cumulative impacts of individual, unrelated developments across the landscape. The designation of critical habitat is an important first step in formulating a landscape approach that considers occupied range, unoccupied range north of the river, and the dispersal corridors that connect them. Designation of critical habitat would also be helpful in evaluating the potential impacts of individual developments on the entire range of the panther. The need for critical panther habitat has been voiced for nearly a decade (Logan et al. 1994; Maehr 1997b), but now that we have a better understanding of dispersal dynamics in this population, it is perfectly justifiable to include the dispersal pathways that subadults are most likely to use (Meegan and Maehr 2002). Whereas such habitat may not be used on a day-to-day basis, it is no less important in restoring normal demographics in this small population. It is likely that critical habitat that includes dispersal pathways would overlap neatly with the Florida Ecological Network (Hoctor et al. 2000; Maehr et al. 2002) and provide additional justification for aggressive land conservation in the area.

The translocation of female panthers to south-central Florida should occur without delay. The current arrangement of landscape barriers may not be enough to prevent a motivated male from making the journey, but it appears to be an absolute barrier to females. Not only would resident females be magnets for future dispersing males, but subsequent reproduction (or lack thereof) would provide direct evidence regarding the suitability of this landscape for the subspecies. Successful reestablishment of a population in south-central Florida would become an important conservation umbrella that would benefit other species as well as the programs (such as Florida Forever) that are intended to protect all of Florida's biodiversity. A failure for translocated females to reproduce successfully despite opportunities to mate should trigger large-scale habitat restoration so that the landscape might once again support a breeding population.

Finally, recovery agencies must adopt a vision of the panther outside the confines of south Florida. To do otherwise is to accept a future zoo population that is maintained like a domesticated herd. Thus, the biological needs of the panther must come before individual vanity and the geocentric biases of public land managers. We also have a moral obligation to promote panther recovery.

> At least part of the answer is that it cannot be ethically acceptable to absolve persons from moral responsibility for situations—like the panther's being endangered—that they have created. . . . This notion that we have a debt of reparation or compensation, because of what persons have done to the Florida panther, seems plausible in part because loss of the subspecies has and will hurt the human and biotic community in Florida, and hurt it in a variety of scientific, cultural, symbolic, aesthetic, and economic ways. (Shrader-Frechette and McCoy 1993)

Conclusion: Recovery and the Dispersal Phenomenon

A few generalities and lessons from the study of Florida panther dispersal may be helpful to others engaged in recovering populations of large carnivores:

1. Large carnivores are inherently programmed to seek out and colonize vacant range.
2. Population density is a determinant in dispersal effort and distance moved.
3. Females, especially among solitary species, may be a particular challenge because of their philopatric tendencies, but even they should be expected to disperse long distances on occasion.
4. Managers must not only be aware of the biological and landscape challenges in restoring large carnivores but must also be effective in eliminating the cultural and bureaucratic barriers that can be more problematic.
5. Private land owners must be partners (not adversaries) in restoring populations of large carnivores.
6. Where artificial barriers limit movements of dispersers, or where only males appear proficient at crossing them, translocation of females should be attempted.
7. Increasing the probability for success among dispersing individuals will facilitate population growth and recovery.

Finally, the re-creation of viable dispersal pathways, based on the documented landscape wanderings of home range-seeking panthers, will go far in relieving some of the moral debt that we have accrued in more than a century of landscape dislocation and denaturement.

Acknowledgments

I appreciate the ongoing efforts of field workers in documenting the consequences of recovery efforts on the Florida panther. E. D. Land, D. B. Shindle, and M. A. Orlando were particularly helpful in the development of the ideas in this chapter.

Literature Cited

Adams, L. W., and L. E. Dove. 1989. *Wildlife Reserves and Corridors in the Urban Environment: A Guide to Ecological Landscape Planning and Resource Conservation.* National Institute for Urban Wildlife, Columbia, MD.

Alberto, D. 2003. "Numbers grow; panthers head to Hillsborough," *Tampa Tribune,* March 16.

Alvarez, K. 1993. *Twilight of the Panther: Biology, Bureaucracy and Failure in an Endangered Species Program.* Myakka River Publishing, Sarasota, FL.

Belden, R. C., and B. W. Hagedorn. 1993. "Feasibility of translocating panthers into northern Florida," *Journal of Wildlife Management* 57:388–397.

Clobert, J., E. Danchin, A. A. Dhondt, and J. D. Nichols, eds. 2001. *Dispersal,* Oxford University Press, New York.

Gray, J. 1948. "Migration of vertebrate animals," in *The Soil and the Sea,* ed. T. Williams, 181–193. Saturn Press, London.

Griffith, B., J. M. Scott, J. W. Carpenter, and C. Reed. 1989. "Translocation as a species management tool: Status and strategy," *Science* 245:477–480.

Hanski, I. 1999. *Metapopulation Ecology.* Oxford University Press, Oxford.

Hansson, L. 1991. "Dispersal and connectivity in metapopulations," *Biological Journal of the Linnean Society* 42:89–103.

Harris, L. D., and J. Scheck. 1991. "From implications to applications: The dispersal corridor principle applied to the conservation of biological diversity," in *Nature Conservation 2: The Role of Corridors,* ed. D. A. Saunders and R. J. Hobbs, 191–220. Surrey Beatty and Sons, Chipping Norton, NSW.

Hoctor, T. S., M. H. Carr, and P. D. Zwick. 2000. "Identifying a linked reserve system using a regional landscape approach: The Florida ecological network," *Conservation Biology* 14:984–1000.

Howard, W. E. 1960. "Innate and environmental dispersal of individual vertebrates," *American Midland Naturalist* 63:152–161.

Ims, R. A., and D. O. Hjermann. 2001. "Condition-dependent dispersal," in *Dispersal,* ed. J. Clobert, E. Danchin, A. A. Dhondt, and J. D. Nichols, 203–216. Oxford University Press, New York.

Lidicker, W. Z. Jr., and R. L. Caldwell, eds. 1982. *Dispersal and Migration. Benchmark Papers in Ecology,* vol. 11. Hutchinson and Ross, Stroudsburg, PA.

Logan, K. A., and L. L. Sweanor. 2001. *Desert Puma: Evolutionary Ecology and Conservation of an Enduring Carnivore.* Island Press, Washington, DC.

Logan, T., A. C. Eller Jr., R. Morrell, D. Ruffner, and J. Sewell. 1994. *Florida Panther Habitat Preservation Plan*. Florida Panther Interagency Committee, U.S. Fish and Wildlife Service, Gainesville.

MacDonald, D. W., and D. D. P. Johnson. 2001. "Dispersal in theory and practice: Consequences for conservation biology," in *Dispersal*, ed. J. Clobert, E. Danchin, A. A. Dhondt, and J. D. Nichols, 358–372. Oxford University Press, New York.

Maehr, D. S. 1990. "The Florida panther and private lands," *Conservation Biology* 4:167–170.

———. 1997a. "Comparative ecology of bobcat, black bear, and Florida panther in south Florida," *Bulletin of the Florida Museum of Natural History* 40:1–176.

———. 1997b. "The Florida panther and the Endangered Species Act of 1973," *Environmental and Urban Issues* 24:1–8.

———. 1997c. *The Florida Panther: Life and Death of a Vanishing Carnivore*. Island Press, Washington, DC.

———. 2001. "Large mammal restoration: Too real to be possible?" in *Large Mammal Restoration: Ecological and Sociological Challenges in the 21st Century*, ed. D. S. Maehr, R. F. Noss, and J. L. Larkin, 345–354. Island Press, Washington, DC.

Maehr, D. S. and G. B. Caddick. 1995. "Demographics and genetic introgression in the Florida panther," *Conservation Biology* 9:1295–1298.

Maehr, D. S., and J. P. Deason. 2002. "Wide-ranging carnivores and development permits: Constructing a multi-scale model to evaluate impacts on the Florida panther," *Clean Technologies and Environmental Policy* 3:398–406.

Maehr, D. S., and R. C. Lacy. 2002. "Avoiding the lurking pitfalls in Florida panther recovery," *Wildlife Society Bulletin* 30:971–978.

Maehr, D. S., R. Grimes, and J. L. Larkin. 1999. "Initiating elk restoration: The Kentucky case study," *Proceedings of the Annual Conference of the Southeastern Association of Fish and Wildlife Agencies* 53:350–363.

Maehr, D. S., E. D. Land, and M. E. Roelke. 1991a. "Mortality patterns of panthers in southwest Florida," *Proceedings of the Annual Conference of Southeastern Fish and Wildlife Agencies* 45:201–207.

Maehr, D. S., E. D. Land, and J. C. Roof. 1991b. "Social ecology of Florida panthers," *National Geographic Research and Exploration* 7:414–431.

Maehr, D. S., E. D. Land, D. B. Shindle, O. L. Bass, and T. S. Hoctor. 2002. "Florida panther dispersal and conservation," *Biological Conservation* 106:187–197.

McCally, D. 1999. *The Everglades: An Environmental History*. University Press of Florida, Gainesville.

Meegan, R. P., and D. S. Maehr. 2002. "Landscape conservation and regional planning for the Florida panther," *Southeastern Naturalist* 1:217–232.

Noss, R. F., M. A. O'Connell, and D. D. Murphy. 1997. *The Science of Conservation Planning*. Island Press, Washington, DC.

Orlando, M. A. 2003. "The ecology and behavior of an isolated black bear population in west central Florida." M.S. thesis, University of Kentucky, Lexington.

Randall, D., W. Burggren, R. Eckert, and K. French. 2002. *Animal Physiology: Mechanisms and Adaptations*. W. H. Freeman, New York.

Rusmore, B., A. Swaney, and A. D. Spader, eds. 1982. *Private Options: Tools and Concepts for Land Conservation*. Island Press, Washington, DC.

Shrader-Frechette, K. S., and E. D. McCoy. 1993. *Method in Ecology: Strategies for Conservation*. Cambridge University Press, Cambridge, U.K.

Simpson, G. 1940. "Mammals and land bridges," *Journal of the Washington Academy of Sciences* 30:137–163.

Stanley Price, M. R. 1989. "Reconstructing ecosystems," in *Conservation for the Twenty-First Century,* ed. D. Western and M. Pearl, 210–218. Oxford University Press, New York.

Swingland, I. R., and P. J. Greenwood. 1983. *The Ecology of Animal Movement.* Oxford University Press, New York.

U. S. Fish and Wildlife Service (USFWS). 1987. "Florida Panther (*Felis concolor coryi*) Recovery Plan." Prepared by the Florida Panther Interagency Committee for the U. S. Fish and Wildlife Service, Atlanta, GA.

Wiens, J. A. 2001. "The landscape context of dispersal," in *Dispersal,* ed. J. Clobert, E. Danchin, A. A. Dhondt, and J. D. Nichols, 96–109. Oxford University Press, New York.

Yalden, D. W. 1993. "The problems of reintroducing carnivores," in *Mammals as Predators,* ed. N. Dunstone and M. L. Gorman, 289–306. Oxford University Press, New York.

CHAPTER 10

State Wildlife Governance and Carnivore Conservation

Martin Nie

The reintroduction and natural recovery of wolves (*Canis lupus*) in the United States is a familiar story. Their reintroduction into Yellowstone National Park and central Idaho in the mid-1990s could fairly be called one of the most controversial and closely watched environmental stories in American history. Less than 10 years later, wolves have successfully recolonized a large portion of the northern Rockies region of Montana, Idaho, and Wyoming. Farther east, in the Lake Superior region, wolves have spread from western Ontario and northern Minnesota into northern Wisconsin and the Upper Peninsula of Michigan. In other regions, the story is not so far along. In the Southwest, the reintroduction of Mexican wolves (*Canis lupus baileyi*) has faced challenges, with human-caused wolf mortality and vehement protest by the livestock industry major factors. Studies of the biological, social, and political feasibility of reintroducing wolves into northern New England and the southern Rockies are still ongoing. And natural wolf dispersal from the northern Rockies into northern Utah, eastern Oregon, and Washington has sparked renewed debate over wolves and their management.

Less well known is what exactly happens after successful reintroduction and natural recovery. Generally, after a species is successfully recovered under the Endangered Species Act (ESA), it is delisted, and management responsibility reverts from the U.S. Fish and Wildlife Service (USFWS) to the states, once they can demonstrate that the species is no longer at risk and will not become so in the future. In short, wolves in the northern Rockies and Lake Superior regions will no longer be protected by the ESA and will be managed differently by state wildlife agencies and various American Indian tribes. What will not change is the political conflict and controversy surrounding wolves. As discussed below, conflict will likely increase and become even more divisive in the future.

This chapter examines the next stage of wolf politics and offers suggestions on how conflict might be dealt with in the future. It begins by reviewing the significant sociopolitical dimensions of this value- and interest-based political conflict. It then reviews state wildlife governance—the dominant institutions, processes, and funding sources used by most state wildlife agencies. Next, it looks to Alaska for "policy lessons" on conflict and state wolf management. Alaska provides a number of important lessons in what to do and what not to do in the future when it comes to state wolf management, because wolves have never been federally protected there. The chapter finishes by reviewing the public participation and funding opportunities and challenges presented to those states in the northern Rockies region.

The major conclusion is that the dominant state wildlife management paradigm, including the role played by state wildlife commissions and game-dependent agency budgets, may play a significant role in future political conflict if change is not forthcoming. The question of how decisions are made will be central. The most controversial wolf management decisions of the future will be value- and ethics-based. The chapter therefore recommends that agencies engage in a form of "authentic public participation," be it stakeholder based or otherwise. If controversial management decisions are made by a state wildlife commission that is not reflective of the public's myriad values toward wildlife, we should not be surprised to see public attempts at gaining a voice through the ballot initiative process.

Background: Dimensions of the Wolf Debate

Before examining the political conflicts on the horizon, it is first important to recognize that wolf recovery and management is a value- and interest-based political conflict transcending science, biology, and technical approaches to problem solving. In many respects, the debate is quite similar to that of other value-laden political disputes, such as abortion and the death penalty. Sometimes, as with the case of livestock depredation problems, differences may stem more from economic interests than from fundamental value differences. It is constructive to frame the debate in terms of values and interests, partly because many future debates over wolves will have very little to do with science and technical wildlife management. Instead, many of them will revolve around competing human values and different constructions of the natural world. Political decision makers, interest groups, and wildlife managers hoping for a technical fix to these conflicts—a "bio bullet," if you will—are setting themselves up for disappointment and frustration.

The debate also goes beyond wolves, because a number of "symbol and surrogate issues" are involved and raise the stakes of the conflict. As noted by Steven Primm and Tim Clark of the Northern Rockies Conservation Cooperative:

> Conservation of large carnivores is also linked inextricably to several broader issues. The U.S. Endangered Species Act (ESA), ecosystem management, use of western public lands, and other contentious issues all manifest themselves in the issue of carnivore conservation. Wrangling over carnivore conservation is also often a "surrogate" for broader cultural conflicts: preservation versus use of resources, recreation-based economies versus extraction-dependent economies, urban versus rural values, and states' rights versus federalism. (Primm and Clark 1996)

Issues and debates over land use and ecosystem management, the contested role of the federal government in reintroducing wolves; the role of conservation biology, wilderness preservation, and large-scale land conservation strategies; the implementation and future of the ESA; tribal participation and management authority; and the future of rural communities make the political debate over wolf recovery and management even more controversial and acrimonious (Nie 2003). Sometimes, because of the nature of the policy problem, it is difficult to disentangle these issues. For instance, large carnivore conservation is inextricably linked to public lands ranching. But in other cases, wolves are consciously used by political actors as a springboard to advance other political goals and objectives, whether it be additional wilderness designation or the reworking of the ESA.

State Wildlife Governance: The Big Institutional Picture

Essential to mapping the possible future of wolf politics and policy are the political institutions and policymaking processes used in state wildlife management. This section examines two of the most important—state wildlife commissions and the ballot initiative process—and the relationship between them. Many conservationists and wildlife advocates believe that nongame interests are not adequately represented on state wildlife commissions. Thus the ballot initiative process is sometimes used as an alternative way to make controversial wildlife policy decisions. Both have played contested roles in the wolf politics of Alaska, and they may, if we learn nothing from this case, play the same role in the lower forty-eight states in the near future.

State Wildlife Commissions

State wildlife commissions were created in the 1930s by sport hunters and conservationists to protect wildlife from widespread market hunting. As a way to institutionalize such protection, and safeguard wildlife, sport hunters were placed on commissions to adopt and enforce wildlife laws. This is only one example of how sport hunters have played an important part in America's conservation history (Reiger 1975). Forty-seven states have some sort of formal citizen board, commission, or advisory council that either makes, recommends, or advises on fish, wildlife, park, and natural resource management decisions.

Most states have a commission, board, or council dealing with either fish and wildlife or natural resources (Musgrave and Stein 1993). Members are typically appointed by a governor and subject to state legislative approval. Most states also have requirements for commission membership, such as a general knowledge of wildlife issues, and political and geographic balance (Musgrave and Stein 1993). Some go even further. Prior to a ballot initiative in 1996, for example, Massachusetts had a statutory requirement that five of seven fisheries and wildlife board members hold a sporting license (Minnis 1998; Pacelle 1998). In Montana, at least one member must be experienced in the breeding and management of domestic livestock, and in 2003 the Montana legislature introduced legislation that would mandate additional commission composition requirements.

The policy and management responsibilities of state wildlife commissions and boards vary, from setting fish and game seasons and bag limits to charting broader management goals and objectives. In Alaska, the Alaska Department of Fish and Game (ADFG) is responsible for implementing policies approved by the state's Board of Game (BOG), a seven-member body whose members are appointed by the governor and confirmed by the state legislature. Although having no fiscal authority, the BOG has constitutional and statutory authority to set game policies, make harvest and management regulations, and regulate subsistence use of wildlife on state and private lands (National Research Council 1997).

State wildlife agencies were also created as a way to protect wildlife and manage game species on a sustained-yield basis. A classic client-manager relationship has historically characterized state wildlife management (Decker et al. 1996). Wildlife managers saw their primary responsibility as providing resources (fish and game) to their clients—anglers, hunters, and trappers. This management paradigm was characterized by agencies being "captured" by these interests, and this capture is partially explained by how these agencies are funded. A large percentage of funding for state wildlife management comes directly from the sale of hunting and fishing licenses, or indirectly from federal assistance funds generated from excise taxes on fishing and hunting equipment. In Alaska, for example, approximately 90% of ADFG Division of Wildlife Conservation's funding is provided by hunters and trappers through license and tag fees and matching federal funds derived from hunting-related excise taxes (ADFG 2001).

And Alaska is not an anomaly. In Wyoming, no general funds or state tax dollars are used to fund the Wyoming Game and Fish Department (WGFD). License fees, nonresident application fees, and conservation stamps provide 67% of its revenue; 25% comes from federal aid and grants, much of it related to hunting and fishing equipment excise taxes (WGFD 1998). Consequently, agencies most often direct their resources toward the management of game species. One study, for instance, shows state expenditures for nongame species amounting to roughly 3% of the amount spent on game species (Mangun 1986).

The state wildlife management paradigm characterized by clientism and agency capture was challenged in the 1970s. Changes in the political landscape brought environmental values to the forefront, and these interests had new laws and procedural opportunities to have their voices heard (Decker et al. 1996). An increasing number of Americans were expressing values toward wildlife that went beyond the utilitarian and consumptive (Kellert 1996). The more exclusive traditional "subgovernment" characterizing state game management had been seriously challenged by nonconsumptive interests that did not hunt, fish, or trap (Tober 1989; Mangun et al. 1992). Of course, the state wildlife management paradigm is not the only paradigm to be challenged in recent years. The principles of ecosystem management, conservation biology, and the emergence of stakeholder-based collaborative conservation, among other things, have forced a number of natural resource agencies to take stock and reassess how they make decisions (Cortner and Moote 1999).

Nonconsumptive interest groups have challenged the clientism paradigm by striking at what they see as the root of the problem: the wildlife policymaking *process*. The Humane Society of the United States (HSUS), clearly prioritizing this issue, summarizes: "The 94 percent of Americans who do not hunt are effectively excluded from wildlife management decisions and policy development. Though non-hunters are increasingly knocking on the doors behind which these decisions are made, the states are fighting as never before to keep them closed" (Hagood 1997). HSUS, among other critics, has a number of general complaints about the commission framework, including that (1) it runs counter to the idea of the public interest in wildlife, because most Americans do not hunt, fish, or trap; (2) it favors consumptive values and interests; (3) commission members often have a conflict of interest because of their business interests in the consumptive use of wildlife, such as being guides, outfitters, and taxidermists; (4) nongame species and their habitat requirements are not prioritized; and (5) the public has the right to debate the values, ethics, and morality of various hunting practices (Hagood 1997; Pacelle 1998).

A few interest groups are beginning to challenge the wildlife commission process. Many wildlife advocacy groups in Alaska have been critical of the state's BOG. The Alaska Wildlife Alliance, together with other state-based groups, filed an unsuccessful lawsuit challenging the legality of the "all hunter-trapper monopoly on the BOG" (Alaska Wildlife Alliance 2000–2001). The lawsuit charged that the BOG fails to fulfill its statutory requirement to provide diversity of interest and points of view in its membership and that it violates the state's constitution, which mandates that the state manage wildlife in the public interest. According to the alliance, "When it comes to deciding wildlife issues one might suppose that all sides would be represented in proportion to their occurrence among the Alaskan population. They are not. On state lands the consumptive interests dominate to the near exclusion of other wildlife values" (Joslin 1998).

Criticism has also come from those inside the wildlife professional subculture. Bruce Gill (1996) of the Colorado Division of Wildlife believes that wildlife agencies have been slow to embrace the human dimensions of wildlife management. Similar to the professional history of American forestry, the wildlife profession has been characterized by a utilitarian technical-rational philosophy, with an emphasis on wise use, sustained yield, and scientific expert management. As such, it has been slow to respond to changing American values toward wildlife. Gill (1996) is also critical of how most state wildlife agencies continue to be funded: "Though perhaps unintentional, license fees effectively married public servants to special interests. It was an unholy marriage because it blurred the essential distinction between public interest and special interest and inevitably eroded both scientific credibility and public trust."

Commissions have also been criticized by hunters and fishers who believe they are dominated by large landowners and/or ranchers. In January 2002, for example, the director of the Idaho Department of Fish and Game, Rod Sando, stepped down because of top-down political pressure (Barker and Phillips 2002a). Many interests in the state, from hunters to nonhunting environmentalists, believe that the pressure came from ranchers and farmers upset with Sando's predator control policies, among other issues. This pressure was channeled, they believe, through the governor and onto the state wildlife commission (a seven-member board appointed by the governor and approved by the state senate).

To some, the resignation was seen as undue political meddling and evidence of a "hijacked" wildlife commission that is unnecessarily politicizing state wildlife management (Barker and Phillips 2002b). The Idaho Wildlife Federation, along with other groups, responded with an increasingly common tool in wildlife management: they drafted a ballot initiative to give the public more control over wildlife decisions.

Ballot Initiatives

The state wildlife commission process is one important reason why so many groups are now using ballot initiatives as a way to make wildlife policy and management decisions. Many states adopted the ballot initiative and referendum process between 1898 and 1918, partially because of perceptions of corrupt state legislatures dominated by corporate interests. Populists and progressives, like many conservationists today, wanted more direct participation in the democratic process. The cure for the ills of representative democracy, they argued, was more democracy (Cronin 1989). Twenty-four states (including almost every state in the wolf-inhabited West) allow for some form of the initiative (direct or indirect), whereby citizens can place issues on the ballot if they meet certain criteria such as petition requirements (Initiative and Referendum Institute 2001b). The process continues to be seen by many supporters as an important check on unresponsive institutions (Initiative and Referendum Institute 2001b).

An unprecedented number of wildlife-related citizen ballot initiatives have been voted on in recent years. The Initiative and Referendum Institute's database documents that 30 wildlife-related ballot measures have been voted on from 1916 through 2001 (Initiative and Referendum Institute 2001a). Of these, 19 have occurred since 1990 (Minnis 1998). Wayne Pacelle (1998) of the HSUS finds that voters have sided with animal protection advocates in 10 of 13 ballot initiatives addressing specific hunting and trapping practices since 1990. These initiatives have targeted a number of predator issues, including mountain lion hunting in California, Oregon, and Washington; types of bear hunting in Colorado, Massachusetts, Oregon, and Washington; trapping in Arizona, Colorado, and Massachusetts; bobcat hunting with dogs in Massachusetts and Washington; lynx hunting with dogs in Washington; and aerial "land and shoot" hunting of wolves in Alaska, among others (Williamson 1998). Most wildlife ballot initiatives have focused on specific hunting practices and have not sought to ban all hunting. Initiatives have also focused on the decision-making process and funding issues whereby some portion of sales tax revenue is used for wildlife management (Arkansas and Missouri) (Williamson 1998).

A full review of the case for and against the ballot initiative process is beyond the scope of this chapter (see Zisk 1987; Cronin 1989; Broder 2001). Nevertheless, it is important to quickly review the challenges and recognize the limitations of the initiative process as a way to make wildlife policy and management decisions. A number of factors should be considered here, including:

1. The deliberation, meaningful dialogue, common ground, and possible compromise that is precluded by this type of adversarial and dichotomous (yes/no, for/against) policymaking. This type of zero-sum approach is not only divisive but often leads to the oversimplification of issues, and the vilification of those who refuse to fit into one of two sides.
2. The quality and stability of public opinion concerning wildlife-related issues, including the amount of knowledge voters have about these issues (Manfredo et al. 1997).
3. The role that science, biology and professional wildlife management plays, or doesn't play in the process.
4. The influence of special-interest groups and large sums of money, including the ability of well-financed groups to use political advertising (fairly or unfairly) to frame issues (Broder 2001).
5. The possibility of a "tyranny of the majority" type of situation, in

which urban residents dominate. This process can lead to intrastate conflict and division, pitting urban versus rural residents (Deblinger et al. 1999).
6. The disproportionate effects of wildlife policy and management decisions. Rural citizens are often more personally affected by wildlife decisions than those living in urban places.

Also important to point out is the political backlash to "ballot box biology." The spate of wildlife ballot measures is related to recent action by state legislatures and interest groups seeking to protect the right to hunt and fish in the form of state constitutional amendments. Six states have amended their state constitutions to do so since 1996, and hunters in at least thirteen states are planning to do the same (Jonsson 2002).

Alaska Policy Lessons

The American system of federalism provides an opportunity to look at states as "laboratories of democracy" providing valuable policy lessons in what works, what doesn't work, and why. Alaska provides such an opportunity regarding state wolf management because it is the only state in the nation that has always had state wolf management authority. The state has dealt with a number of controversial wildlife issues using the ballot initiative. Voters in 1996 used the initiative process to ban the controversial practice of same-day airborne land-and-shoot wolf hunting (the practice of flying over wolf habitat, landing a plane near a wolf pack, and shooting wolves once a hunter moves at least 100 yards from the plane). The initiative also prohibited the state from using aircraft in government wolf control programs except in the case of a biological emergency. In 1998 "Alaskans Against Snaring Wolves" were unsuccessful in passing a proposition that would, among other things, prohibit the use of snares in trapping wolves. And in 2000, Alaskans voted to reinstate the ban on public same-day airborne wolf hunting (after the state legislature brought the issue back on the agenda).

A number of prominent wildlife advocates and interest groups in Alaska emphasize the BOG decision-making process as an important reason why the state has had so many wolf-related ballot initiatives. These groups and individuals argue that their values, and the nonconsumptive values of most Alaskans, are not adequately represented on the state's BOG and are thus not taken seriously. The alternative, they insist, is to take these issues to the public via the initiative process. In short, many interests view the ballot initiative as the policymaking venue of last resort. What option is left, they ask, if commissions and state legislatures are not responsive to what they believe are the values and policy preferences of the majority. And such unresponsiveness is often explained by the political power of consumptive and commodity-oriented special interests in state legislatures.

Is there anything that states in the northern Rockies and Lake Superior regions can learn from the Alaskan experience? This question was posed to biologists, wildlife managers, political representatives, hunters and trappers, tribal representatives, wolf advocates, and many others in south-central and interior Alaska during interviews and research for the Wolf Policy Project (an examination of the political dimensions of wolf recovery and management and how states might move forward with carnivore conservation in the future) (Nie 2003). My interpretation of some of the most important lessons is summarized in the following sections.

1. This conflict is value- and interest-based and has multiple sociopolitical dimensions.

A surprising number of stakeholders in Alaska compared the wolf debate to the abortion issue. It boils down to competing values, they insist. "Symbol and surrogate" issues pertaining to such things as subsistence and federal land ownership and regulations further compound this debate and make it more difficult to resolve.

2. We should recognize the limitations of science and biology as a way to resolve wolf-centered political conflict.

Environmental politics, including the subject of wolves, is often characterized by an adversarial form of analysis in which opposing groups use "their science" to forward their policy objectives. If I only had a dollar for every time someone involved in wolf politics told me, "The other side isn't using good science." Even when stakeholders agree on the science, they often filter this science using disparate value systems. And when they focus their attention on the same facts and evidence, they often give them different interpretations (Schön and Rein 1994). We might agree, for example, that more grizzlies are being spotted in more places, but is that because of an increase in their population or because they are moving out of increasingly degraded core habitats?

Of course, science can help answer these questions, but it will also raise new ones that will again be subject to different political interpretations. Furthermore, stakeholders will often call for more biological study, and their opponents will see this as a strategic way to postpone a political or management decision (in some respects, similar to the debate over global climate change). Political representatives often ask scientists to do the impossible, partly because it absolves them of making difficult policy choices. Biology will and must play an important role in the future, but we must not ask or expect biologists to resolve these divisive conflicts and moral disagreements for the public at large. That would be an unfair assignment.

3. The most volatile political conflict is yet to come.

Future questions and issues concerning wolf management and control will be even more controversial than those pertaining to wolf reintroduction. Wolf conflict in Alaska has usually centered on questions pertaining to wolf control: if and when it is necessary, for whom is it being done, and how it should be implemented. This level of conflict, moreover, has sustained itself without any significant public lands ranching in the state. Many wolf advocates and others, inside and outside Alaska, are troubled by the methods in which wolves have been managed in the past. The 1996, 1998, and 2000 ballot initiatives were the result of such protest. Where will wolves be allowed? From where, and for what reason, will they be removed or killed? For what reason will they be controlled? Will wolves be hunted or trapped by the public? How? Future conflict will also likely go beyond environmentalists and ranchers, in part, because the welfare of individual wolves will be at stake. This means that animal rights and welfare groups may play a larger and more vocal role in the future, and many of these groups focus on *individual* animal welfare, not only on a viable wolf *population*.

4. Debate will often center on questions pertaining to the effect of wolves on game and for what purpose wildlife is being managed.

Not surprisingly, the debate in Alaska revolves around complex biological questions about predator-prey dynamics (National Research Council 1997). Simplified, hunters and other consumptive users of wildlife argue that in conjunction with other factors (e.g., severe winters, bears, human harvest levels), wolves can keep moose and caribou populations at unacceptable low levels, from a biological and social standpoint. These users argue that wolves must therefore be managed and controlled, and control means killing wolves. Wolf advocates in the state often reject the very premise on which these arguments are based (Haber 1996). Many contend that hunters and trappers, including the Board of Game (BOG), want the state's wildlife managed as a game farm. That is, they want the state's moose and caribou populations maintained at unnaturally high levels, and the only way to do this is by keeping wolves at unnaturally low levels.

5. Process and inclusion matters and is an important part of wolf-centered political conflict.

The important role that the BOG plays in setting wildlife policy in the state is often criticized by environmentalists and wolf advocates. It is a clear example, they say, of single-use wildlife management made in an exclusive process by a select number of consumptive interests and values. They contend that the BOG, dominated by hunters, fishers, and trappers, does not reflect the wildlife values of most Alaskans. As a consequence, they see any meaningful policy change regarding wolves and wildlife as an uphill battle unless the policy process is reconfigured in some way. Perceptions of the BOG as being exclusive and unreflective also helps explain the 1996 and 2000 wolf initiatives. According to some wolf advocates, they were essentially forced to use the process because of the way most wildlife policy decisions are made. If their values are not represented on the BOG, and these values are perceived as being in the majority, then the obvious route is to take their case to the public in the form of a ballot initiative.

6. A stakeholder-based collaborative approach to future wolf conflict can be successful.

ADFG is now engaged in various stakeholder-based approaches to wolf and wildlife management. "Stakeholding" provides no magic bullet. Its limitations and weaknesses are obvious, from questions of who is a stakeholder to what role and decision-making authority they should be given. But it is often no worse, and sometimes much better, than other ways of making controversial wildlife management decisions. If done properly, the stakeholder approach provides a more inclusive and proactive way of making value- and interest-based political decisions. It is also possible to use a stakeholder approach within a larger democratic and legislative framework. The Fortymile Caribou Herd Management Planning Team, for example, utilized a stakeholder-based approach to a controversial caribou recovery and wolf control problem within a larger institutional framework, thus providing multiple levels of accountability and numerous checks and balances (see Todd 2002). In other words, decisions made by the stakeholder team went through existing institutions and processes, including the state legislature, the governor, and the BOG.

It would be advantageous for states like Minnesota, Wisconsin, Michigan, Montana, Idaho, and Wyoming either to start assembling stakeholder teams or to maintain existing ones that can proactively deal with future issues and likely scenarios regarding state wolf management. It would be worthwhile to deal with these issues before the crisis and trainwreck stage of decision making. The use of stakeholders in wolf management will also become less problematic because the ESA will no longer complicate things. That is, as long as wolf viability is guaranteed, states will have some flexibility in how they manage wolves, and how they go about making these controversial decisions.

Moving Forward with Wolf Management

Future debate over wolf management will have explicit moral and ethical dimensions. Take the following questions and cases for example:

- In Wisconsin, hunters hunting black bear with hounds are compensated by the state when their dogs are killed by wolves on public land (see Naughton-Treves et al. 2003). Game farm operators are also compensated when they lose game to wolves. Is such compensation socially and financially acceptable?
- Should ranchers losing livestock to wolves on public lands (and in wilderness areas) be compensated the same as those losing livestock on private property?
- How much micromanagement and human manipulation of wolves is tolerable? When does management become outright control, and when does control become morally objectionable?
- If wolf population control is agreed on, how should it be done? The possible methods—whether vasectomy or tubal ligation, relocation, trapping, snaring, "denning," a public harvest, or aerial hunting—will most definitely matter to the public at large, and environmental, wildlife advocacy, and animal rights and welfare groups.

The "right" answer to these and other questions will not come from the expert, in whatever guise, from biologist to policy analyst. Instead, the answer should come out of a political process that is as open, fair, and inclusive as possible. Because we live in a pluralistic democratic society, we will never solve the "wolf problem" once and for all. Answers "good enough" today may be unacceptable tomorrow.

Understanding that these sorts of questions will be central in the future, it is important to find decision-making processes that encourage open and constructive political debate. If, however, these decisions are made through a process that privileges some values over others, or is widely perceived as such, we should not be surprised to find unhappy interests using the ballot initiative process in some wolf states where it is available. So where do we go from here? At the time of this writing, state wolf management planning and the wolf delisting process is ongoing. Nevertheless, Idaho, Wyoming, and Montana provide three approaches to decision making and ways in which future conflict might be managed in the future—for better or worse (see Nie 2003 for how Alaska, Minnesota, and Wisconsin made these decisions). These cases show that the Alaska experience may not be unique, and that Montana has gone the furthest in applying its lessons.

Idaho's Wolf Conservation and Management Plan

Idaho's Wolf Conservation and Management Plan (Idaho 2002) is premised on a problem definition focusing on private property and states' rights (see Clark 1997 for the importance of problem definition in wildlife management). First, the official position of Idaho (adopted as House Joint Memorial No. 5 in 2001) is for the federal government to remove wolves from the state. Given the unlikelihood of this happening, the plan begins by outlining the state constitutional parameters of carnivore management, particularly the inalienable right of "acquiring, possessing and protecting property." This rather general management plan was prepared by the Idaho Legislative Oversight Committee and aims to bring wolves back under commission and Idaho Department of Fish and Game (IDFG) control as either a big-game animal, furbearer, or special classification of predator that provides for a controlled take after delisting.

As with other fish and wildlife issues in the state, the Idaho state legislature has played a contested role in crafting management goals and guidelines. This legislative role is politically interpreted in a number of different ways. For some, it is inappropriate political meddling and micromanagement of wildlife, whereas for others, it is the democratic responsibility of an elected legislature. Does the Idaho legislature adequately represent the values and interests of state residents? Does the legislative committee–interest group decision-making process? How people answer these questions will determine their position on this legislative role. Because of the value and ethics-based nature of the debate, one could make the case that the legislature's intense involvement is the epitome of public participation in wildlife management—participation by the state's elected political representatives. Others see it differently and argue that the legislative commission framework should be either replaced or supplemented with more diverse participatory and/or stakeholder-based processes.

Wyoming's Draft Gray Wolf Management Plan 2003

The Wyoming Game and Fish Commission caused a great deal of controversy in how it proposed to manage wolves following ESA delisting. Because of the gray wolf's classification as a predatory animal in Wyoming statute, the Wyoming Game and Fish Department (WGFD) sought legislation that would change this classification to a "dual status" (WGFD 2003). According to the department's management plan, wolves would be managed as trophy game animals in national parks (Yellowstone and Grand Teton) and wilderness areas (on the Shoshone and Bridger Teton National Forests) in the northwestern part of the state. In wilderness areas, a regulated public take (hunting and trapping) would be used for wolf population management. Outside of these designated areas, however, wolves would statutorily be classified as a predatory animal that could be killed at any time (an unregulated public take).

Unsurprisingly, WGFD's proposal to list wolves as predatory in most of the state was not well received by many interests, including the Montana Governor's Office, which had lobbied WGFD to amend their draft management plan (WGFD 2002). Delisting wolves in the northern Rockies region requires Idaho, Montana, and Wyoming to have state wolf management plans that are acceptable to the USFWS. One unacceptable plan, by one state, means no delisting for all three states. They are tied at the hip. Thus, Montana and Idaho would not gain management authority until Wyoming came up with a plan acceptable to the USFWS.

The Wyoming Game and Fish Department recognizes the important questions and controversy likely to emerge from state wolf management and "acknowledges the complexity of the political, social and environmental factors associated with wolves and their management" (WGFD 2003). As such, the department undertook a public participation strategy that is perhaps best characterized as an "inquisitive approach" (Chase et al.

2000). In this approach to stakeholder involvement, agencies invite input from stakeholders, in the form of public meetings, open houses, scientific polling surveys, and other methods. As researchers in Cornell's Human Dimensions Research Unit explain, "Human dimensions inquiry can augment the information base with systematic research about stakeholders' beliefs, attitudes, preferences, expectations, and behaviors" (Chase et al. 2000). But this form of public involvement has its limitations as well, for while it provides valuable input, "[I]t does not tell agencies what they 'ought to do' in a situation where stakeholder values are the determining factor for management decisions" (Chase et al. 2000).

The WGFD used a number of inquisitive approaches to finding out what the public thought about its draft wolf management plan, including scoping sessions, open houses, written public comment, and a commissioned public opinion survey of state residents (Kruckenberg and Burkett 2003). Nevertheless, some citizens were quite unsatisfied with the public participation process, complaining, among other things, about the department's unusually short (35 days) public comment period on the draft plan, and the fact that they were being asked to comment on a plan that was widely reported and understood as being unacceptable to the USFWS. In other words, they were being asked to comment on a proposal in a state of flux and one with a very high degree of uncertainty. The question "Why bother?" was most certainly asked by many whose concerns were probably well-founded, considering that the final management plan maintained the same dual status provisions as the draft (WGFD 2002, 2003). But some learning may have also taken place. Alaska was threatened with tourism boycotts over its wolf management practices (Hampton 1997), and WGFD notes from its public participation assessment that "a surprising number of these comments indicated that the individual would both boycott and encourage others to boycott Wyoming's tourism industry if this plan were pursued" (Kruckenberg and Burkett 2003).

The Wyoming process raises a number of important questions about the natural resource and wildlife decision-making process in general, and the inquisitive approach to stakeholder involvement in particular. In assessing public comment and opinion, for example, what "public" and what form of public participation matters most—and does it matter at all? What happens in this "black box" between public inputs and decisions outputs? What should happen? The WGFD even seemed somewhat perplexed by this issue because the results from the statewide telephone survey contradicted the results from the analysis of written comments. It notes that "the written comments from residents indicated very high levels of both support and opposition to dual classification, but the survey indicated a majority of residents would support such a plan" (Kruckenberg and Burkett 2003). Such incongruity stems in part from the different constituencies providing input: those answering the telephone survey are passive respondents, while those choosing to participate represent a more distinct and active segment of the society. But perhaps the most important question is what role the commission ought to play vis-à-vis this public input. Such input is basically meaningless if the state wildlife commission had already made up its mind.

Montana's Wolf Conservation and Management Plan

Montana's plan is based on a different institutional and participatory arrangement and is widely considered the most comprehensive and representative plan in the region. It is founded on a general appreciation of the significant sociopolitical dimensions of wolf management. The Montana Wolf Conservation and Management Plan is largely based on the advice and recommendations of a diverse 12-member citizens group, the Montana Wolf Management Advisory Council, and an Interagency Technical Committee (MFWP 2002). Montana Fish, Wildlife and Parks "intends to honor the diverse perspectives and interests of our citizens and the national public [and] the state will consider a spectrum of interests in designing and implementing a balanced, responsive program that recognizes the opportunities and addresses the challenges faced by people directly affected by wolves" (MFWP 2002).

This planning process and stakeholder design might best be thought of as a way of "embedding collaboration." The advisory council, in other words, is but one part of a larger process and more inclusive participatory framework. With scientific guidance by the technical committee, the advisory council forwarded a set of agreed-upon principles to guide future state wolf management, and these principles were then drafted into a more comprehensive and detailed management plan. Montana's Environmental Policy Act (MEA) also requires state agencies to go through the environmental impact statement (EIS) process when proposed actions are of this significance. MFWP handled these multiple responsibilities by using the advisory council's recommendations as its preferred alternative in the draft state EIS and recommending it in the final EIS (MFWP 2003).

Thus, state residents and the national public had an opportunity to provide input, through public comment for example, on the alternatives, including that put forth by the citizen's group. Note that this stakeholder team was not given unchecked decision-making authority, nor were they given a role that necessarily challenged the state's wildlife commission. The commission, the governor, and the director of MFWP have important veto powers (as they do in other states as well). The legislature also remains important because it must approve the agency's proposed budget and Montana law classifying wolves as predators needed to be changed so that it was not an impediment to delisting.

Embedding a stakeholder-based approach in a larger institutional and decision-making framework might alleviate some critics' concerns about accountability. In other words, existing institutional arrangements are not being completely abandoned in the name of stakeholding; rather, they are being supplemented with increased public participation. Such embedding, however, will inevitably lead to complaints of "process fatigue." Similar to U.S. Forest Service complaints of "the process predicament" and "analysis paralysis," critics see this type of stakeholding as simply another layer of process slowing things down and preventing progress. Questions may also arise over using the council's recommendation as MFWP's preferred alternative, for it raises some question about the EIS process in general.

The Funding Dilemma

If Alaska is any guide, future conflict over wolf management will be driven in part by perceptions of agency capture due to game-based budgetary incentives. Some interests in Alaska question any predator control decision made by ADFG because they believe such decisions are based on economic interest, not good biology or public values and opinion. In other words, for whom, and for what purpose are wolves being managed? Language in some state wolf plans foreshadows similar conflict. Idaho's wolf plan, for example, states: "The wolf population will be managed at recovery levels that will ensure viable, self-sustaining populations until it can be established that wolves in increasing numbers will not adversely affect big game populations, the economic viability of IDFG, outfitters and guides, and others who depend on a viable population of big game animals." (Idaho 2002)

But the funding issue goes beyond immediate budgetary considerations related to lost hunter-generated revenue. In Wyoming, larger questions are related to who should pay for carnivore management in general. "Wolves are of national interest, and the national public, not just the license-buying public of Wyoming, should share funding the management of the species" says the Wyoming Game and Fish Department (WGFD 2002). As such, Wyoming and other states have advocated establishing and funding the proposed "Northern Rocky Mountain Grizzly Bear and Gray Wolf Management Trust." As envisioned, "the Trust would originate from a one-time Congressional appropriation and form the basis of an inviolate corpus, upon which the available annual interest would be sufficient to offset most of the three states' cost of managing grizzly bears and wolves" (WGFD 2002). WGFD (2002) advocates the trust approach as a way that "would allow the American public to share in the cost of these management programs, rather than having it fall entirely to the states, which rely almost exclusively on license fees and excise taxes on sporting equipment to support agency programs."

Finding a dependable source of nongame funding is of critical importance. Keep in mind that the reintroduction of wolves into places like Yellowstone was framed by advocates as being in the national interest, be it to restore ecological integrity or for posterity purposes, among other arguments. But the trust proposal also raises the question of what strings, if any, these states will be willing to accept if federal money comes along. If the trust arrangement happens, and Congress decides to give in the name of managing species in the national interest, will it then intervene in controversial management decisions?

Conclusion

Future decisions over wolves and their management will be tough going, however they are made. There is something about this animal that makes finding common ground easier said than done. But at the very least, questions about how these decisions are made and then funded should not compound these difficulties. The public must be engaged in a serious way. "Authentic participation is deep and continuous involvement in administrative processes with the potential of all involved to have an effect on the situation. . . . requires that administrators focus on both process and outcome. . . . [It] means that the public is part of the deliberation process from issue framing to decision making" (King et al. 1998). If not authentic, we should not be surprised to find the Alaska experience being played out in the lower forty-eight. With this new voice comes new responsibilities as well. The public, whether through a trust arrangement or by some other funding mechanism, must be given an opportunity to help pay for carnivore and nongame restoration and management.

Acknowledgments

I would like to thank those individuals interviewed for this paper, and personnel in the Alaska Department of Fish and Game; the Wyoming Game and Fish Department; Montana Fish, Wildlife and Parks; and the Idaho Governor's Office of Species Conservation for answering questions and providing information. A version of the state wildlife governance section of this chapter will appear in *Public Administration Review* (Nie 2004). I wish to thank this journal for permission to republish this section. Readers are encouraged to see this article for possible options and alternatives to the commission or ballot initiative way of making controversial wildlife management decisions.

Literature Cited

Alaska Department of Fish and Game (ADFG), Division of Wildlife Conservation. 2001. Available at www.wildlife.alaska.gov/division_info/overwiew.cfm. Accessed April 9, 2004.

Alaska Wildlife Alliance. 2000–2001. "Hunter/Trapper Monopoly," *The Spirit* 19:1–4.

Barker, R., and R. Phillips. 2002a. "Meetings took place in week before Sando quit," *Idaho Statesman*, February 1:1.

———. 2002b. "Predator politics gets ugly in Idaho," *High Country News*, February 18:3.

Broder, D. 2001. *Democracy Derailed: Initiative Campaigns and the Power of Money*. Harvest Books, New York.

Chase, L. C., T. M. Schusler, and D. J. Decker. 2000. "Innovations in stakeholder involvement: What's the next step?" *Wildlife Society Bulletin* 28:208–217.

Clark, T. W. 1997. *Averting Extinction: Reconstructing Endangered Species Recovery*. Yale University Press, New Haven.

Cortner, H. J., and M. A. Moote. 1999. *The Politics of Ecosystem Management*. Island Press, Washington, DC.

Cronin, T. E. 1989. *Direct Democracy: The Politics of Initiative, Referendum, and Recall*. Harvard University Press, Cambridge, MA.

Deblinger, R. D., W. A. Woytek, and R. R. Zwick. 1999. "Demographics of voting on the 1996 Massachusetts ballot referendum," *Human Dimensions of Wildlife* 4:40–55.

Decker, D. J., C. C. Krueger, R. A. Baer Jr., B. A. Knuth, and M. E. Richmond. 1996. "From clients to stakeholders: A philosophical shift for fish and wildlife management," *Human Dimensions of Wildlife* 1:70–82.

10. State Wildlife Governance and Carnivore Conservation 217

Gill, B. 1996. "The wildlife professional subculture: The case of the crazy aunt," *Human Dimensions of Wildlife* 1:60–69.

Haber, G. C. 1996. "Biological, conservation, and ethical implications of exploiting and controlling wolves," *Conservation Biology* 10:1068–1081.

Hagood, S. 1997. *State Wildlife Management: The Pervasive Influence of Hunters, Hunting, Culture and Money*. Humane Society of the United States, Washington, DC.

Hampton, B. 1997. *The Great American Wolf*. Henry Holt, New York.

Idaho Legislative Wolf Oversight Committee. 2002. *Idaho Wolf Conservation and Management Plan*, as amended by the 56th Idaho Legislature, 2nd Regular Session, March 2002.

Initiative and Referendum Institute. 2001a. *Statewide Database*. Available at www.iandrinstitute.org/statewide_i&r.htm. Accessed April 9, 2004.

———. 2001b. *I & R Factsheet*. Available at www.iandrinstitute.org. Accessed November 19, 2001.

Jonsson, P. 2002. "'Right to hunt' vs. animal rights: What's fair game?" *Christian Science Monitor*. April 2. Available at www.csmonitor.com/2002/0403/p01s04-ussc.htm. Accessed April 8, 2002.

Joslin, P. 1998. "The need for balance," *The Spirit* 17:8.

Kellert, S. R. 1996. *The Value of Life: Biological Diversity and Human Society*. Island Press. Washington, DC.

King, C. S., K. M. Feltey, and B. O. Susel. 1998. "The question of participation: Toward authentic public participation in public administration," *Public Administration Review* 58:317–326.

Kruckenberg, L. L., and C. Burkett. 2003. *Special Report—Draft Wyoming Gray Wolf Management Plan*. Wyoming Game and Fish Department, Cheyenne.

Manfredo, M. J., D. C. Fulton, and C. L. Pierce. 1997. "Understanding voter behavior on wildlife ballot initiatives: Colorado's trapping amendment," *Human Dimensions of Wildlife* 2:22–39.

Mangun, J. C., J. T. O'Leary, and W. R. Mangun. 1992. "Non-consumptive Wildlife-Associated Recreation in the United States: Identity and Dimension," in *American Fish and Wildlife Policy: The Human Dimension*, ed. W. R. Mangun, 175–200. Southern Illinois University Press, Carbondale.

Mangun, W. R. 1986. "Fiscal constraints to nongame management programs," in *Management of Nongame Wildlife in the Midwest: A Developing Art*, ed. J. B. Hale, L. B. Best, and R. L. Clawson, 23–32. Bookcrafters (for the North Central Section of The Wildlife Society), Chelsea, MI.

Minnis, D. L. 1998. "Wildlife policy-making by the electorate: An overview of citizen-sponsored ballot measures on hunting and trapping," *Wildlife Society Bulletin* 26:75–83.

Montana Fish, Wildlife and Parks. 2002. *Montana Wolf Conservation and Management Planning Document (Draft)*. Available at www.fwp.state.mt.us/news/show.aspx?id=2323. Accessed April 9, 2004.

———. 2003. *Montana Gray Wolf Conservation and Management Plan: Final Environmental Impact Statement*. Available at www/fwp.state.mt.us/wildthings/wolf/finaleis/finalwolfeis.asp. Accessed April 9, 2004.

Musgrave, R. S., and M. A. Stein. 1993. *State Wildlife Laws Handbook*. Government Institutes, Inc., Rockville, MD.

National Research Council. 1997. *Wolves, Bears, and Their Prey in Alaska: Biological and Social Challenges in Wildlife Management*. National Academy Press, Washington, DC.

Naughton-Treves, L., R. Grossberg, and A. Treves. 2003. "Paying for tolerance: The impact of depredation and compensation payments on rural citizens' attitudes toward wolves," *Conservation Biology*. 17:1500–1511.

Nie, M. 2003. *Beyond Wolves: The Politics of Wolf Recovery and Management*. University of Minnesota Press, Minneapolis.

———. 2004. "State wildlife policy and management: The scope and bias of political conflict," Public Administration Review 64:206–218.

Pacelle, W. 1998. "Forging a new wildlife management paradigm: Integrating animal protection values," *Human Dimensions of Wildlife* 3:42–50.

Primm, S. A., and T. W. Clark. 1996. "Making sense of the policy process for carnivore conservation," *Conservation Biology* 10:1036–1045.

Reiger, J. F. 1975. *American Sportsmen and the Origins of Conservation*. Oregon State University Printing. 3rd edition, December 2000.

Schön, D. A., and M. Rein. 1994. *Frame Reflection: Toward the Resolution of Intractable Policy Controversies*. Basic Books, New York.

Tober, J. A. 1989. *Wildlife and the Public Interest: Nonprofit Organizations and Federal Wildlife Policy*. Praeger, New York.

Todd, S. 2002. "Building consensus on divisive issues: A case study of the Yukon wolf management team," *Environmental Impact Assessment Review* 22:655–684.

Williamson, S. J. 1998. "Origins, history, and current use of ballot initiatives in wildlife management," *Human Dimensions of Wildlife* 3:51–59.

Wyoming Game and Fish Department (WGFD). 1998. "Financial Status Briefing Statement." Available at: http://gf.state.wy.us/admin/briefs/brief_98/finstate.asp. Accessed April 7, 2004.

———. 2002. "Draft Wyoming Gray Wolf Management Plan." http://gf.state.wy.us/wildlife/wildlife_management/draft_wolf.asp

———. 2003. "Final Wyoming Gray Wolf Management Plan." Available at http://gf.state.wy.us/downloads/pdf/WolfPlanFinal8-6-03.pdf. Accessed April 9, 2004.

Zisk, B. H. 1987. Money, Media and the Grassroots: State Ballot Issues and the Electoral Process. Sage: Newbury Park, CA.

CHAPTER 11

Conserving Mountain Lions in a Changing Landscape

Christopher M. Papouchis

Top carnivores play a vital ecological function by maintaining the integrity and stability of ecosystems (Berger 1999; Terborgh et al. 1999; Miller et al. 2002). Yet at the dawn of the twenty-first century, wide expanses of the North American landscape remain bereft of their full complement of top carnivores. In the western United States, mountain lions (*Puma concolor*) remain the only top carnivore sustaining viable populations throughout the landscape (Diamond 1993; Logan and Sweanor 2001). The wide-ranging mountain lion is considered an important focal species for efforts to conserve native ecosystems and biodiversity at the landscape level (Beier 1993; Logan and Sweanor 2001). Indeed, the species' extensive distribution throughout the Americas, its adaptability to a wide array of habitats, and its superior predatory ability may afford it a more important ecological role than any other top predator in the Western Hemisphere (Murphy et al. 1999). Conserving self-sustaining populations of mountain lions where they exist and facilitating their recovery in historic range offer significant benefits for the preservation of healthy ecosystems and biodiversity.

The management of mountain lions, however, remains driven by a traditional philosophy that emphasizes the utility of mountain lions to humans (Murphy et al. 1999; Torres et al., in prep.) despite their ecological importance; mounting public opposition to some traditional management practices (Torres et al. 1996; Teel et al. 2002); and a lack of sound biological information on mountain lion populations (Logan and Sweanor 2001). Biologically, human understanding of mountain lions remains nascent because of the inherent difficulty and expense of studying this elusive and solitary creature. Furthermore, a dearth of accurate and reliable scientific information on the size and vigor of mountain lion populations cloaks management of the species in a veil of uncertainty (Logan and Sweanor 2001). To further complicate matters, mountain lions, like other large carnivores, elicit powerful and often negative emotions in humans (Kellert et al. 1996). Even when scientific information is available, mountain lion management can be inordinately dictated by traditional attitudes and local politics (Miller et al. 2002; Laundré and Clark 2003), to the detriment of lion populations.

Still, mountain lions have managed to survive in western North America despite efforts to eradicate them, and they appear poised to recolonize parts of their historic range in the East. At the same time, the sociopolitical and ecological landscape continues to change. Public attitudes toward the species have improved in many areas, but rapidly increasing human populations, development and road building in mountain lion habitat, and a rise in the killing of mountain lions for sport and predator control pose a significant threat to the viability of local populations and the integrity of regional metapopulations (Logan and Sweanor 2001). Addressing the complexities and uncertainties inherent in conserving this large, wide-ranging carnivore will require scientists, wildlife agencies, and conservationists to reform current management policies and develop regional and collaborative strategies.

A History of Persecution

Mountain lions have roamed the Western Hemisphere for at least 100,000 years, with a distribution that exceeds that of any other terrestrial wild mammal in the Americas (Sunquist and Sunquist 2002). Once they ranged from coast to coast, from the Yukon to the southern tip of Chile, but they have experienced significant reductions in their range and numbers since European colonization (Hansen 1992). Fear and hatred dominated the perceptions of European settlers toward the large carnivores—mountain lions, grizzly bears, and wolves—as they considered them predators of their livestock, competitors for the species they hunted for food, and threats to their safety (Kellert et al. 1996). The contempt for large carnivores, born largely from misunderstanding and myth (Acuff 1988), led to the eradication of wolves (*Canis lupus*), grizzly bears (*Ursus arctos*), and mountain lions from vast expanses of the continent.

The extinction of mountain lions from North America was generally considered a *fait accompli* as bounty programs and the formidable resources of the federal government were engaged to eradicate mountain lions throughout the West in the first decades of the twentieth century. Statements by such an influential conservationist as President Theodore Roosevelt that the mountain lion was "the destroyer of the deer, the lord of stealthy murder, facing his doom with a heart both craven and cruel" (Roosevelt 1913) reinforced the perception that the best mountain lion was a dead mountain lion. Even the National Geographic Society remarked, "Unfortunately the predatory habits of this splendid cat are such that it cannot continue to occupy the same territory as civilized man and so is destined to disappear before him" (Nelson 1918). From 1907 to 1977, at least 66,655 mountain lions were killed in the western United States (Nowak 1976), and more likely died from scavenging poison-laced carcasses (Ken Logan, Colorado Division of Wildlife, pers. comm.). However, though wolves and grizzlies were extirpated from most of the West, the enigmatic and retiring mountain lion endured, surviving in rugged, mountainous areas (Young and Goldman 1946).

In the face of this onslaught there were those who endeavored to present a more accurate and less sensationalized image of the species. Committees of the Ecological Society of America, which included such ecological luminaries as Aldo Leopold and Ernest Seton, began lobbying for the termination of carnivore eradication programs and for the protection of wildlife habitat in the late 1920s (Soulé and Noss 1988). Joseph Grinnell et al. (1937) concluded that the mountain lion was the most misunderstood mammal in California, and Frank Hibben (1939) wrote that "the common everyday generalities which have been established in lay and scientific minds alike have, perhaps, denied the predator his true place." Still, it would take nearly half a century for wildlife agencies to begin to incorporate into management policies the concept that mountain lions and other predators are invaluable components of the natural world.

The first long-term research on the species was not initiated until the 1960s, when Dr. Maurice Hornocker and his associates began a seminal investigation into a mountain lion population in the remote mountains of Idaho (Logan and Sweanor 2001). They pioneered the use of radiotelemetry to provide the first detailed observations of lion population dynamics, their social organization, and their relationship with prey (Hornocker 1969, 1970; Seidensticker et al. 1973). Hornocker's research helped lay to rest many of the myths that had dominated human perceptions of mountain lions and augmented the growing public and scientific appreciation for the mountain lion as an integral component of natural ecosystems. While Colorado and Nevada had eliminated bounties and afforded mountain lions limited protection by classifying them as game animals by 1965 (Nowak 1976), it was not until after Hornocker had published his findings that the other western states began to follow suit (K. Logan, pers. comm.). Whereas previously mountain lions had been considered vermin, they now were seen as a valuable "resource."

The Status of Mountain Lions in the United States and Canada

Today, mountain lions occupy only 50% of their historic range in the hemisphere (Logan and Sweanor 2000), and roughly 33% of their U.S. range (Grigione 2000). Arizona, California, Colorado, Idaho, Montana, New Mexico, Oregon, Utah, Texas, Washington, Wyoming, and the Canadian provinces of Alberta and British Columbia appear to support viable populations of mountain lions (Padley 1996; Harveson et al. 2003). Wildlife agencies develop estimates of the size and trend of mountain lion populations using a variety of tools, including hunting data, frequency of conflicts and sightings, and track surveys. However, these methodologies are considered by many scientists and conservationists and even some state agencies to be inaccurate and unreliable (Logan and Sweanor 2001). Ascertaining the true size and vigor of mountain lion populations remains an elusive goal.

Mountain lions appear to be recolonizing parts of their former range in the eastern United States and Canada (Walker 2003). Mountain lion populations have begun to recover in South Dakota, and confirmed sightings in several Great Plains and Great Lakes states suggest that mountain lions are dispersing from populations in the Rocky Mountains and southern Canada. Rusz (2001) reported the presence of a breeding population in Michigan. While most sightings of lions in the eastern United States are likely misidentifications or just wishful thinking (Cardoza and Langlois 2002), credible sightings are increasing in frequency (Bolgiano et al. 2000), stirring renewed and vigorous debate on the status and future of mountain lions in the East (Cardoza and Langlois 2002; Maehr et al., in press). At least two conservation groups (the Eastern Cougar Foundation and the Eastern Cougar Network) have formed in the last decade to document sightings and other signs and to address the related issues. Researchers are still trying to determine the status of cats sighted in the East (i.e., wild or escaped/released captive animals; Bolgiano et al. 2000).

Mountain lions in Florida (USFWS 1987) and southern California (Beier 1993; Sauvajot and Riley 2002) are presently jeopardized by habitat loss and fragmentation and unsustainable levels of mortality due to contact with humans. In Florida, mountain lions, also known as Florida panthers (P.c. coryi), have been listed as a federally endangered species since 1967 (USFWS 1987), and while infusing the population with transplants from Texas has staved off extinction to this point, intense human development will likely prevent this population from regaining its former vigor (see Maehr, Chapter 9 of this volume).

Changing Attitudes Toward Mountain Lions

Previously, negative public attitudes toward mountain lions led to state and federal government efforts to eradicate mountain lions and other carnivores. However, current attitudes toward carnivores make such management strategies less acceptable (see Breck, Chapter 1 of this volume). To be sure, there are still those who favor wholesale reductions or even complete removal of mountain lion populations (Etling 2001), but, overall, public perception of mountain lions and other carnivores has improved, and opposition to traditional management programs has increased (Kellert et al. 1996). Residents of the Pacific Coast states have afforded mountain lions the most protection: California voters banned the sport hunting of mountain lions in 1990 and rejected efforts to reinstate it in 1996; Oregon and Washington voters prohibited the use of hounds to hunt mountain lions in 1994 and 1996, respectively.

Surveys conducted in the Intermountain West have also found positive attitudes toward mountain lions. In a 2001 survey of the northern Rocky Mountain states of Idaho, Montana, and Wyoming, a majority of respondents from the general public indicated that mountain lions were an important component of the ecosystems they occupy and that people had a responsibility to learn to coexist with them (Montag et al. 2003). Similarly, Riley and Decker (2000) found that a majority of survey respondents in Montana considered the presence of mountain lions to be a sign of a healthy environment and as benefiting their overall quality of life. In Utah, researchers found that 50% of urban residents and 38% of rural residents disapproved of mountain lion hunting, while a majority of each disapproved of using hounds to hunt mountain lions (Teel et al. 2002). There was also little support in Utah for killing mountain lions to "boost" game species (Krannich and Teel 1999).

The Wyoming Game and Fish Department surveyed Wyoming residents in 1995 and found that though nearly 50% of respondents thought that mountain lion hunting should continue, 57% opposed the use of hounds to hunt mountain lions, and 80% agreed that females with kittens should be protected from hunting (Gasson and Moody 1995). A 2001 survey conducted in Arizona, Colorado, and New Mexico found that citizens considered mountain lions to be the species best representative of the southern Rockies heritage and landscape (Decision Research 2001). Further, roughly 50% of respondents understood that mountain lions, wolves, and grizzly bears play an important ecological role. The survey also found that nearly 66% would support a ban on hound hunting of mountain lions (Decision Research 2001).

Contemporary attitudes toward mountain lions would appear to bode well for the survival of the species. However, public sentiment remains highly variable and capricious (Murphy et al. 1999), and negative views held by a minority still hold sway over management decisions (Clark et al. 1996). For instance, because of an apparent rise in reported human–mountain lion conflicts, some rural residents in northeastern Washington argued that mountain lions were overabundant. They created political pressure sufficient to force the Washington Department of Fish and Wildlife to issue several mountain lion hunting tags, despite protest by the agency's own biologists (R. Spencer, WDFW, pers. comm.). As escalating human immigration, development, and activity in mountain lion habitat continues to increase the likelihood of contact between humans and mountain lions, efforts to conserve mountain lions will become increasingly complicated and divisive (Riley and Decker 2000).

Threats to Mountain Lion Populations

Human activity remains the primary threat to the long-term viability of mountain lion populations and metapopulations (Logan and Sweanor 2001). The loss, degradation, and fragmentation of habitat is considered the principal threat (Murphy et al. 1999; Logan and Sweanor 2001), though excessive human-caused mortality, or overkill, can also negatively impact populations (Logan and Sweanor 2001), particularly those that are small and isolated (Beier 1993). As human densities in the United States increase, so does the probability that mountain lion populations will be driven to extinction (Woodroffe 2000; but see Linnell et al. 2001).

The human population in the West has exploded in the last half century, growing from 14 million in 1940 to 60 million in 2004 (U.S. Census Bureau 2001). The West is outpacing the rest of the country in population growth: 13 of the 20 fastest-growing counties in the United States are located in the intermountain states of Colorado, Utah, Idaho, and Nevada (Center of the American West 2003). By 2025, it is projected that more than 85 million people will live in the American West (U.S. Census Bureau 2001). Rapidly expanding human development has exacerbated habitat loss and fragmentation and increased negative interactions between cougars and livestock, and cougars and humans, intensifying the intentional and unintentional killing of mountain lions (Sweanor et al. 2000). Nowell and Jackson (1996) found that the disappearance of large felids outside of protected areas begins with the loss and fragmentation of their habitat, followed by their direct persecution by humans as the result of conflicts with domestic animals or people.

As the western landscape continues to be transformed by humans, lions have been forced to adapt in order to survive (Baron 1999). The vast array of western public lands and undeveloped private lands has provided core habitat for mountain lions and enabled their survival to the present. However, many remote habitats that once served as core reserves for mountain lions are now compromised; less than 6 to 9% of the mountain lion's total range lies within protected areas, many of them too small to even enclose the home range of a single mountain lion (Nowell and Jackson 1996). Mountain lions that find their home ranges under residential or industrial development must relocate and contest with neighboring individuals for habitat or attempt to survive in their present area. Similarly, without sufficient wild habitat and corridors, dispersing subadults seeking a territory may encounter residential areas. Roadkill can become a significant source of mortality (Beier 1993). People who generally express positive attitudes toward mountain lions may be less tolerant of mountain lions that frequent areas near their homes (Riley and Decker 2000).

Attacks on humans, though exceedingly rare, may further increase public concern and create negative attitudes toward mountain lions (Torres et al. 1996). Beier (1991, 1992) documented 43 confirmed mountain lion attacks on humans in North America between 1890 and 1990, 11 of them fatal. In the years since, at least 9 other people have been killed. These events have been well publicized in the media and in books (e.g., Etling 2001; Deurbrouck and Miller 2002; Baron 2003), often in a sensationalized fashion that makes the likelihood of attacks seem greater than it is. Aune (1991) and Baron (2003) have argued that lions that live near developed areas may become habituated to humans and thus more likely to attack people, which could account for the increase in attacks in recent years. However, Beier (1992) found no substantial evidence to support this hypothesis. Furthermore, Ken Logan (pers. comm.) has argued that humans, by killing those individual mountain lions that prey on domestic animals and threaten public safety, may be selecting for lions that tend to avoid humans. Notwithstanding the apparent increase in attacks, mountain lions pose a far lesser threat than most perils faced by humans. For example, a person is nearly 100,000 times more likely to be killed while driving a car than to be killed by a mountain lion.

Current Management Trends

The management of mountain lions primarily involves the regulation of mortality (Murphy et al. 1999). The principal reasons mountain lions are killed are for sport, for preying on domestic animals, for competing with hunters for game, for research, and for posing a threat to public safety. Currently, the rate of mountain lion kills is the highest ever recorded (Figure 11.1; Torres et al., in prep.). In the western United States at least 3,500 mountain lions were reported to have been killed by humans during 2002, according to wildlife agency data. Torres et al. (in prep.) found that at least 53,000 mountain lions had been killed in the western United States since 1970.

The vast majority of human-caused mountain lion mortality in the West is caused by sport hunting, the principal tool used by wildlife agencies to "manage" mountain lion populations (Logan and Sweanor 2001). As of 2003, Arizona, Colorado, Idaho, Montana, New Mexico, Oregon, Utah, Washington, and Wyoming had regulated hunting seasons for mountain lions. Mountain lions in Texas are classified as a predatory species and have no protections (Hansen 1992), while those in California have not been hunted since 1972, when Governor Ronald Reagan signed a legislative moratorium. California mountain lions are presently classified as a special-

11. Conserving Mountain Lions in a Changing Landscape 227

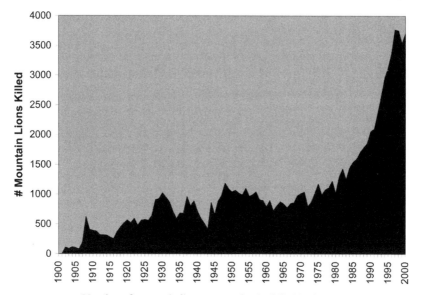

FIGURE 11.1 Number of mountain lions reported to be killed in the western United States. Information represents all kill data recorded during the 1900s. Up through the 1960s the records are almost exclusively bounty records. From this period through 1999, the kill data are primarily regulated hunting, and to a much lesser extent depredation.

This information needs to be interpreted in the context that it includes only recorded mortalities, and different time periods have different potential biases. These biases primarily include underreporting from missing information, and unrecorded/unknown kills.

During the bounty period there was a paid incentive to report kills. However, some scientists have suggested that widespread and poorly documented poisoning programs may have been an additional significant mortality factor. The latter part of the century recorded regulated hunting kills, but likely was underreported with respect to depredation kills.

Similar to widespread poisoning programs that no longer exist, vehicle mortality is now a potentially significant mortality factor, for which there are no records. Therefore, these data conservatively represent total kill data for the last century, primarily biased by unrecorded mortalities. (Torres et al., in prep.).

ly protected mammal, as the result of a ballot initiative approved by the voters in 1990 (Torres et al. 1996). In South Dakota, mountain lions were removed from the state endangered species list and were classified as game mammals in 2003; a hunting season may be proposed within the next few years (G. Vandell, SD Game, Fish and Parks, pers. comm.).

Hunting management strategies are ostensibly designed using the best available science to protect hunted mountain lion populations from overexploitation. However, Logan and Sweanor (2001) argued that mountain lion "hunting management in most western states is a far cry from science."

Most western states—especially the intermountain states of Colorado, Idaho, Montana, Utah, and Wyoming—have steadily increased hunting quotas on mountain lions in the past several decades because of increased hunter interest; perceived increases in lion numbers (Logan and Sweanor 2001); and the belief that mountain lions are depressing ungulate populations (Laundré and Clark 2003), threatening public safety, or predating on domestic animals. Often the assumptions that drive management are scientifically unsupported—for example, reducing predation on livestock is frequently cited as a benefit of sport hunting though evidence is lacking (Sunquist and Sunquist 2002).

Sport hunting can affect mountain lion populations in several ways, though to what extent or duration is poorly understood (Murphy et al. 1999). Researchers in Utah and New Mexico found that after simulating a sport hunt by removing 27% and 47% of mountain lions from their respective study areas, the population did not recover to preremoval levels for 24 and 31 months, respectively (Lindzey et al. 1992; Logan and Sweanor 2001). Hunting can depress populations until the habitat is recolonized by individuals dispersing from surrounding habitat (provided there is connectivity with surrounding populations), creating a sink effect (i.e., neighboring populations are drained of their young lions because they are killed when they disperse into the empty "sink" area) (Lindzey et al. 1992; Sweanor et al. 2000; D. Stoner, pers. comm.). In areas where mountain lion populations are connected, such as in the Southwest, reduction or extinction of mountain lions in one subpopulation may destabilize the metapopulation (Sweanor et al. 2000).

Additionally, by killing or removing adult resident mountain lions, particularly females, sport hunting may affect the dynamics of mountain lion populations and social relations among surviving adults and subadults (Anderson 1983; Murphy et al. 1999). Sport hunting further represents an unnatural selective pressure since hunters typically select for the largest mountain lions and may thus be affecting the genetic heredity of populations (Logan and Sweanor 2001). Hunting may also cull timid individuals that are quick to tree when chased by hunting dogs, leaving more aggressive cats to pass on their genes and thus possibly increasing attacks on humans (Deurbrouck and Miller 2002).

In recent decades, the risk of overhunting populations has increased as the result of more hunters in the field and the increased accessibility of remote areas via roads and the use of all-terrain motorized vehicles and snowmobiles (Laundré and Clark 2003). The number of lions killed by hunters has more than doubled since 1990 (Torres et al., in prep.), and sport hunting mortality may be overstressing lion populations in several states (Logan and Sweanor 2001; Dawn 2002; Laundré and Clark 2003).

However, protecting mountain lions from sport hunting pressure does not make mountain lion populations immune to human impacts (Shaw 1989). Since sport hunting was initially prohibited by the California legislature in 1972, the number of lions killed for preying on domestic livestock and pets has increased and is now the largest reported source of human-caused mountain lion mortality in California, peaking at 149 in 2001 (Torres et al. 1996; Cullens and Papouchis 2003). Although fewer lions are killed in California than in any other western state, development in mountain lion habitat and the attendant mortality due to conflicts with people threaten to jeopardize several regional mountain lion populations (e.g., the western slope of the Sierra Nevada Mountains; S. Torres, CDFG, pers. comm.). As humans continue to alter the natural landscape, conflicts between humans and mountain lions are expected to increase throughout the West (Torres et al. 1996).

Mountain Lion Conservation in the Twenty-First Century

Although there is still serious scientific debate about some aspects of mountain lion ecology, one area that has become incontrovertible is their important ecological role (Maehr et al. 2003). Conservation of mountain lions and their habitat yields profound benefits to the conservation of biodiversity. Terborgh et al. (2001) found that after 7 years of isolation, islands created by a hydroelectric dam in Venezuela that were too small to support mountain lions and jaguars lost nearly 75% of the vertebrate species and were overrun by the remaining herbivorous species to the extent that the canopy trees could not regenerate. In the eastern United States, the absence of mountain lions and gray wolves has led to an overabundance of white-tailed deer, which has significantly impacted native vegetation and many species of birds (McShea et al. 1997). Unfortunately, current management programs for mountain lions seldom reflect the ecological importance of the species (Miller et al. 2002).

Mountain lion conservation and management is rife with scientific uncertainty, a reality that must be recognized within policies that affect the species. Rather than continuing to rely almost exclusively on ecological research to guide conservation efforts, it will be necessary to adopt an interdisciplinary approach that incorporates the fields of conservation biology, sociology, political science, economics, and others (Clark et al. 1996; MacDonald 2001). The long-term conservation of mountain lions alongside human society is attainable, provided management policies are favorable (Linnell et al. 2001).

The present controversy over the management of mountain lions and other large carnivores is emblematic of the changes taking place in society at large, the scientific community, and the wildlife management profession itself (see Nie, Chapter 10 in this volume). Ensuring the long-term viability of mountain lion populations throughout their North American range in the face of a rapidly expanding and urbanizing human population requires the reconsideration of the philosophical and methodological paradigms that have guided management of the species heretofore. Though conservation scientists have argued for wildlife agencies to change their focus from a single-species, utilitarian approach to an ecosystem-level, conservation approach, ingrained philosophies and funding sources have proved a significant impediment to change (Maehr et al. 2001). Torres et al. (in prep.) assert that wildlife agencies must broaden the scope of their puma management plans to include the ecological value of lions as keystone predators and as focal species. Further, conservation policies also should consider the emotional, intellectual, spiritual (Kellert et al. 1996), and intrinsic values (Bekoff 2001) that affect how people perceive this charismatic species. Critically, retaining the invaluable ecological role of mountain lions on the landscape will require agencies to strive to maintain ecologically effective populations of lions across the landscape rather than to continue the present policy of managing for only minimum viable populations.

Devising Landscape-Level Conservation Strategies

Cougar populations often function at larger scales than human administrative units (e.g., states), and therefore conservation efforts must be long-range, integrated, and cooperative across geographic and political boundaries (Beier 1993; Murphy et al. 1999; Sinclair et al. 2001; Papouchis and Cullens 2003; Torres et al., in prep.). State wildlife agencies should facilitate regional management by working with other local, state, and federal agencies to develop standards for the collection of data on mountain lions, identifying mountain lion habitat and linkages, and involving stakeholders in adopting and implementing conservation and management policies (Papouchis and Cullens 2003; Torres et al., in prep.). Presently, a panel of mountain lion scientists and managers is working to address current management deficiencies and is developing a set of clearly enunciated guidelines addressing all facets of cougar conservation (S. Negri, pers. comm.).

Sound management and conservation of mountain lions rely on comprehensive scientific data. Yet the cost of mountain lion research can be prohibitive, and the results of a single study are often difficult, if not impossible, to extrapolate to other areas, given the variability of cougar populations and habitat (Murphy et al. 1999). Developing a sound understanding of changing interactions between lions and people (patterns, hotspots, trends, etc.) depends on the ability to share and analyze similar data (Green 1991) and on maintaining records of mountain lion-human encounters (Logan and Sweanor 2001). Wildlife managers, field researchers, and conservation organizations would benefit from the development of consistent and accessible databases that incorporate data on mountain lion populations, including mortality, human interactions, depredation, natural prey base, and habitat use/availability (Murphy et al. 1999). However, there is currently no long-term multistate repository for mountain lion data, and without a consistent standard, each state often collects and stores data differently. Creating a national standardized database could provide rapid access to data, help enable regional assessments, and facilitate multistate collaboration (Papouchis and Cullens 2003).

Habitat fragmentation has been identified as the primary cause of the current extinction crisis (Wilcox and Murphy 1985) and is a primary threat to the viability of mountain lion populations. Beier (1993) concluded that in order to persist at least 100 years, mountain lion populations that receive several immigrant lions per decade require a minimum of 600–1,600 km^2, while those that are isolated need 1,000–2,200 km^2. To offset the risks of habitat loss and overkill, the designation of large reserve areas where mountain lions would be protected from hunting and control efforts has been proposed by researchers and managers (Logan and Sweanor 2001; Laundré and Clark 2003; Torres et al., in prep.) and at least one state agency (Washington Department of Fish and Wildlife 2002).

Reserves would serve as source populations to outlying habitats where mountain lions would remain subject to management actions. Identifying and protecting mountain lion dispersal and travel corridors that link reserve areas would facilitate the dispersal and movement of individuals, promoting gene flow among subpopulations and maintaining functional metapopulations (Beier 1993; Logan and Sweanor 2001). However, the designation of reserves has faced opposition by some rural communities that may consider them breeding grounds for "marauding lions." A 2002 Washington Department of Fish and Wildlife proposal to identify and set aside from hunting 10% of lion habitat in each management unit was withdrawn after significant opposition from rural communities (D. Martorello, WDFW, pers. comm.).

Developing Collaborative Networks

Collaborative efforts to educate and involve the wide array of stakeholders in policymaking decisions will be integral to creating regional conservation strategies. Agencies must recognize the challenges of conserving mountain lion habitat and self-sustaining populations of mountain lions and align all advocates for wildlife, including both the hunting and the nonhunting public (Torres et al., in prep.). Teel et al. (2002) cautioned that "because the success of many traditional wildlife management strategies is increasingly based on public approval, it is risky for wildlife resource agencies to ignore public sentiment, particularly when it is in the form of opposition."

If 1993 to 2003 is any example, the use of ballot initiatives by the public to alter wildlife management practices will continue unless wildlife agencies and commissions redefine who they consider as constituents to include those who have a more protectionist attitude toward wildlife. Chase et al. (2002) suggested that "agencies should invest the resources necessary to implement stakeholder involvement processes that use scientific information, have genuine influence on decisions, treat citizens fairly, and promote communication and education."

Wildlife agencies, conservation groups, and communities also need to expand efforts to educate the public as to the ecological importance of mountain lions, threats to their vitality, and the risks and responsibilities inherent in living and recreating in mountain lion country. Coordinated statewide efforts are integral, and the best results may be through community forums and workshops that involve and engage citizens. Efforts to inform current residents in mountain lion habitat should involve local community organizations, developers, real estate agents, and chambers of commerce. For example, managers could engage local citizens in mountain lion research, such as track surveys, to help personalize people's relationship to mountain lions and increase their active involvement as stakeholders. Wildlife agencies have a statutory responsibility to inform the public, but collaboration among interested groups is essential.

Encouraging Nonlethal Resolution of Conflicts

Management of conflicts with mountain lions typically involves killing or relocating the individual mountain lion involved, or conducting habitat-level population reductions through sport hunting or predator control programs (Ruth et al. 1998). Rather than continuing to rely on lethal control to address livestock depredation and other human–mountain lion conflict, nonlethal approaches should be employed whenever possible (Hansen 1992; Ross and Jalkotzy 1995; Murphy et al. 1999), and the development of nonlethal techniques should be the focus of concentrated research (Logan and Sweanor 2001).

Capturing and relocating mountain lions that come into conflict with humans may be a viable option in some instances (Ross and Jalkotzy 1995; Linnell et al. 1997; Ruth et al. 1998), and young cougars that are ready to disperse or have dispersed are the best candidates (generally 12–27 months old; Ruth et al. 1998). The Washington Department of Fish and Wildlife is studying the effectiveness of radio collaring and relocating so-called "no-harm/no-foul" cougars that wander into rural and suburban neighborhoods and other areas of dense human habitation but cause no harm (R. Spencer, WDFW, pers. comm.). At present, however, state agencies generally refuse to relocate individual mountain lions that come into conflict with humans, citing liability and cost concerns, among others.

The most effective methods for reducing conflicts between mountain lions and livestock may be those traditional techniques that have been used, in some cases for millennia, with a wide variety of carnivores. Responsible animal husbandry and nonlethal predator aversion techniques are widely recommended by state wildlife agencies. The methods applicable to mountain lions include enclosures, fencing, livestock guardian dogs, clearing brush and landscaping with plants that do not attract deer, installing outdoor lighting, restricting the activity of pets outside (especially at night), using herders to accompany and protect open-range livestock, choosing appropriate livestock, bringing in livestock when lambing or kidding, and raising mixed herds of cattle and sheep. However, although these techniques have been the focus of research for reducing conflicts with other predators such as coyotes, wolves, and bears, little information is available on their effectiveness with mountain lions.

Other nonlethal methods deserving of study include the use of aversive conditioning, such as using specially trained dogs to deter mountain lions from frequenting certain areas (Logan and Sweanor 2001). For example, hounds are used in the eastern Sierra Nevada Mountains in an attempt to deter mountain lions from remaining in the vicinity of areas populated by endangered bighorn sheep. Additional aversive conditioning methods include taste aversion (Nowell and Jackson 1996) and projectiles such as rubber bullets (Baron 2003).

Community-based programs that aim to reduce human–mountain lion conflicts, such as the Living with Lions program developed by the Mountain Lion Foundation, a national conservation and education organization, can also be implemented to engage local residents (Mountain Lion Foundation 2003). Such small-scale, collaborative, nonlethal projects can be highly effective, particularly in areas with a history of conflicts.

Developing Conservation-Related Research

Increasing the biological and ecological understanding of mountain lions is critical to their long-term conservation. Because of the rapidly changing landscape of the West and the inherent threats to mountain lions, research efforts should be focused on how humans influence the viability of carnivores on scales of 100 to 1,000 years (Murphy et al. 1999). Investigations into how human recreational activity affects the movement of mountain lions and the distribution of their prey are needed (Joslin and Youmans 1999). Logan and Sweanor (2001) provided numerous recommendations, including development of relatively inexpensive, but accurate and precise techniques to assess the status of lion populations; investigation of mountain lion behavior in "peopled" landscapes to ascertain their ability to avoid or tolerate human activity and determine ways to repel them; further study of translocated mountain lions and the improvement of mountain lion translocation guidelines; development of aversive conditioning techniques to nonlethally deter mountain lions from remaining in areas of human habitation and centers of activity; and examination of the effectiveness of livestock guardian dogs to protect livestock from mountain lions. Additionally, human dimensions research should aim to further discern public attitudes toward mountain lions and ascertain how they impact management policies.

Most studies of mountain lions have been relatively short-term, several years at most, and provide only a snapshot of mountain lion population dynamics for a particular population at a given point in time. Even longer-term research projects, such as the 10-year study conducted by Ken Logan and Linda Sweanor in New Mexico (Logan and Sweanor 2001), while providing a greater perspective and more utility, are still too short to provide truly comprehensive conclusions. Optimally, studies should span several generations of mountain lions (K. Logan, pers. comm.).

Conclusion

Conservation strategies for mountain lions need to be adaptive and based on scientifically rigorous strategies that seek locally and regionally viable populations on healthy and connected landscapes. They must also take into account the social and political realities of coexisting with a large carnivore that is capable of taking down domestic livestock much larger than itself, competing with hunters for prey, and on very rare occasions even attacking and killing a human. Ultimately, humans will determine whether mountain lions are allowed to persist (Logan and Sweanor 2001). If humans can overcome the historical obstacles and predator prejudice that have prevented us from developing sound, long-term conservation strategies for mountain lions and other keystone carnivores, perhaps the natural heritage of the North American landscape—and human society—can be preserved.

Acknowledgments

This chapter is dedicated to my grandmother, Charlotte Nelson, who passed away during its writing. If all people had her drive to live and learn, this planet would be a far different place. Thanks to Ken Logan, Lynn Sadler, Camilla Fox, and Nicholas Papouchis for their helpful comments on the manuscript and to Rick Hopkins, Susan Morse, David Stoner, and many others for their observations of mountain lion ecology and management.

Literature Cited

Acuff, D. S. 1988. "Perceptions of the mountain lion, 1825–1986, with emphasis on *Felis concolor californica*." M.A. thesis, University of California, Davis.

Anderson, A. E. 1983. *A Critical Review of Literature on Puma* (Felis concolor). Colorado Division of Wildlife, Special Report No. 54.

Aune, K. E. 1991. "Increasing mountain lion populations and human–mountain lion interactions in Montana," in *Mountain Lion–Human Interaction Symposium*, ed. C. S. Braun, 86–94. Department of Natural Resources, Division of Wildlife, Denver, CO.

Baron, D. 1999. "Wild in the suburbs," *Boston Globe Magazine*. August 19.

———. 2003. *The Beast in the Garden: A Modern Parable of Man and Nature*. Norton, New York.

Beier, P. 1991. "Cougar attacks on humans in the United States and Canada," *Wildlife Society Bulletin* 19:403–412.

———. 1992. "Cougar attacks on humans: An update and some further reflections," *Proceedings of the Vertebrate Pest Conference* 13:365–367.

———. 1993. "Determining minimum habitat areas and habitat corridors for cougars," *Conservation Biology* 7:94–108.

Bekoff, M. 2001. "Human-carnivore interactions: Adopting proactive strategies for complex problems," in *Carnivore Conservation*, ed. J. L. Gittleman, S. M. Funk, D. MacDonald, and R. K. Wayne, 179–195. Cambridge University Press, Cambridge.

Berger, J. 1999. "Anthropogenic extinction of top carnivores and interspecific animal behaviour: Implications of the rapid decoupling of a web involving wolves, bears, moose and ravens," *Proceedings of the Royal Society of London, Series B* 266:2261–2267.

Bolgiano, C., T. Lester, D. W. Linzey, and D. S. Maehr. 2000. "Field evidence of cougars in eastern North America," in *Proceedings of the Sixth Mountain Lion Workshop*, ed. L. A. Harveson, P. M. Harveson, and R. W. Adams, 34–39. Austin, TX.

Cardoza, J. E., and S. A. Langlois. 2002. "The eastern cougar: A management failure?" *Wildlife Society Bulletin* 30:265–273.

Center of the American West. 2003. *Development, Land Use, and Population Trends in the American West*. Available online at www.centerwest.org/futures.

Chase, L. C., W. F. Siemer, and D. J. Decker. 2002. "Designing stakeholder involvement strategies to resolve wildlife management controversies," Wildlife Society Bulletin 30:937–950.

Clark, T. W., P. C. Paquet, and A. P. Curlee. 1996. "General lessons and positive trends in large carnivore conservation," *Conservation Biology* 10:1055–1058.

Cullens, L. M., and C. M. Papouchis. 2003. "Community-based conservation of mountain lions." Paper presented at the *Seventh Mountain Lion Workshop*, May 15–17, 2003, Jackson, WY.

Dawn, D. 2002. "Management of cougars (*Puma concolor*) in the Western United States." M.S. thesis, San Jose State University, San Jose, CA.

Decision Research. 2001. *Southern Rockies Wildlife and Wilderness Survey Report*. Decision Research, Eugene, OR.

Deurbrouck, J., and D. Miller. 2002. *Cat Attacks: True Stories and Hard Lessons from Cougar Country*. Sasquatch Books, Seattle, WA.

Diamond, J. 1993. "Cougars and corridors," *Nature* 365:16–17.

Etling, K. 2001. *Cougar Attacks: Encounters of the Worst Kind*. Lyons Press, Guilford, CT.

Gasson, W., and D. Moody. 1995. *Attitudes of Wyoming Residents on Mountain Lion Management*. Wyoming Game and Fish Department, Planning Report No. 40.

11. Conserving Mountain Lions in a Changing Landscape

Green, K. A. 1991. "Development of a data base for analysis of information about lion sightings and lion-human interactions," in *Mountain Lion–Human Interactions,* ed. C. E. Braun. Colorado Division of Wildlife, Denver.

Grigione, M. 2000. "The impacts of wildlife management on cat conservation, using the mountain lion as an example." Presentation at the National Wildlife Federation's Endangered Cats of North America Workshop, Yulee, FL.

Grinnell, J., J. S. Dixon, and J. M. Linsdale. 1937. *Fur-Bearing Mammals of California.* University of California Press, Berkeley and Los Angeles.

Hansen, K. 1992. *Cougar: The American Lion.* Northland Publishing, Flagstaff, AZ.

Harveson, L. A., P. M. Harveson, and R. W. Adams. 2003. *Proceedings of the Sixth Mountain Lion Workshop.* Texas Parks and Wildlife, Austin.

Hibben, F. C. 1939. "The mountain lion and ecology," *Ecology* 20:584–586.

Hornocker, M. G. 1969. "Winter territoriality in mountain lions," *Journal of Wildlife Management* 33:457–464.

———. 1970. "An analysis of mountain lion predation upon mule deer and elk in the Idaho primitive area," *Wildlife Monographs* No. 21.

Joslin, G., and H. Youmans. 1999. *Effects of Recreation on Rocky Mountain Wildlife: A Review for Montana.* Committee on Effects of Recreation on Wildlife, Montana Chapter of the Wildlife Society.

Kellert, S. R., M. Black, C. R. Rush, and A. J. Bath. 1996. "Human culture and large carnivore conservation in North America," *Conservation Biology* 10: 977–990.

Krannich, R. S., and T. L. Teel. 1999. *Attitudes and Opinions About Wildlife Resource Conditions and Management in Utah: Results of a 1998 Statewide General Public and License Purchaser Survey.* Final project report for the Utah Division of Wildlife Resources, Utah State University, Institute for Social Science Research on Natural Resources, Logan.

Laundré, J., and T. W. Clark. 2003. "Managing puma hunting in the western United States through a metapopulation approach," *Animal Conservation* 6:159–170.

Lindzey, F. G., W. D. Van Sickle, S. P. Laing, and C. S. Mecham. 1992. "Cougar population response to manipulation in southern Utah," *Wildlife Society Bulletin* 20:224–227.

Linnell, J. D. C., R. Aanes, J. E. Swenson, J. Odden, and M. E. Smith. 1997. "Translocation of carnivores as a method for managing problem animals: A review," *Biodiversity and Conservation* 6:1245–1257.

Linnell, J. D. C., J. E. Swenson, and R. Andersen. 2001. "Predators and people: Conservation of large carnivores is possible at high human densities if management policy is favourable," *Animal Conservation* 4:345–349.

Logan, K. A., and L. L. Sweanor. 2000. "Puma," in *Ecology and Management of Large Mammals in North America,* ed. S. Demarais and P. R. Krausman, 347–367. Prentice Hall, Upper Saddle River, NJ.

———. 2001. *Desert Puma: Evolutionary Ecology and Conservation of an Enduring Carnivore.* Island Press, Washington, DC.

MacDonald, D. W. 2001. "Postscript—Carnivore conservation: Science, compromise and tough choices," in *Carnivore Conservation,* ed. J. L. Gittleman, S. M. Funk, D. MacDonald, and R. K. Wayne, 524–538. Cambridge University Press, Cambridge.

Maehr, D. S., T. S. Hoctor, and L. D. Harris. 2001. "Remedies for a denatured biota: Restoring landscapes for native carnivores," in *Wildlife, Land, and People: Priorities for the 21st Century,* ed. R. Field, R. J. Warren, H. Okarma, and P. R. Sievert, 123–127. Wildlife Society, Bethesda, MD.

Maehr, D. S., M. J. Kelly, C. Bolgiano, T. Lester, and H. McGinnis. 2003 "Eastern cougar recovery is linked to the Florida panther: Cardoza and Langlois revisited," *Wildlife Society Bulletin.* 31:849–853

McShea, W. J., H. B. Underwood, and J. H. Rappole. 1997. *The Science of Overabundance: Deer Ecology and Population Management.* Smithsonian Institution Press, Washington, DC.

Miller, B., B. Dubelby, D. Foreman, C. Martinez del Rio, R. Noss, M. Phillips, R. Reading, M. E. Soulé, J. Terborgh, and L. Willcox. 2002. "The importance of large carnivores to healthy ecosystems," *Endangered Species Update* 18:202–210.

Montag, J. M., M. E. Patterson, and B. Sutton. 2003. *Political and Social Viability of Predator Compensation Programs in the West.* Draft project report, University of Montana, Missoula.

Mountain Lion Foundation. 2003. "Living with lions protects lions and livestock," *Mountain Lion Foundation Newsletter,* Fall: 1–2.

Murphy, K. M., Ross, P. I., and M. G. Hornocker. 1999. "The ecology of anthropogenic influences on cougars," in *Carnivores in Ecosystems: The Yellowstone Experience,* ed. T. W. Clark, A. P. Curless, S. C. Minta, and P. M. Kareiva, 77–102. Yale University Press, New Haven, CT.

Nelson, E. W. 1918. *Wild Animals of North America.* National Geographic Society, Washington, DC.

Nowak, R. M. 1976. *The Cougar in the United States and Canada.* U.S. Department of the Interior, Fish and Wildlife Service, Washington, DC, and New York Zoological Society.

Nowell, K. and P. Jackson. 1996. *Wild Cats, Status Survey and Conservation Action Plan.* IUCN, Gland, Switzerland.

Padley, W. D., ed. 1996. *Proceedings of the Fifth Mountain Lion Workshop,* February 27–March 1, 1996. The Southern California Chapter of the Wildlife Society, San Diego.

Papouchis, C. M., and M. Cullens. 2003. "Improving our understanding of mountain lion management trends: The value of consistent multi-state record keeping." Paper presented at the Seventh Mountain Lion Workshop, May 15–17, 2003, Jackson, WY.

Riley, S. J., and D. Decker. 2000. "Wildlife stakeholder acceptance capacity for cougars in Montana," *Wildlife Society Bulletin* 28:931–939.

Roosevelt, T. 1913. "Hunt on the rim of the Grand Canyon," *Outlook,* October 4, 259–266.

Ross, P. I., and M. G. Jalkotzy. 1995. "Fates of translocated cougars, *Felis concolor,* in Alberta," *Canadian Field-Naturalist* 109:475–476.

Rusz, P. 2001. *The Cougar in Michigan: Sightings and Related Information.* Michigan Wildlife Habitat Foundation, Bath.

Ruth, T. K., K. A. Logan, L. L. Sweanor, M. G. Hornocker, and L. J. Temple. 1998. "Evaluating cougar translocation in New Mexico," *Journal of Wildlife Management* 62:1264–1265.

Sauvajot, R., and S. P. D. Riley. 2002. "Mountain lions, habitat fragmentation and wildlife movement corridors in urbanizing landscapes of southern California." Paper presented at the Defenders of Wildlife's Carnivores 2002 Conference, Monterey, CA.

Seidensticker, J. C., M. G. Hornocker, W. V. Wiles, and J. P. Messick. 1973. "Mountain lion social organization in the Idaho primitive area," *Wildlife Monographs* 35:1–60.

Shaw, H. 1989. *Soul Among Lions: The Cougar as Peaceful Adversary.* Johnson Publishing, Boulder, CO.

Sinclair, E. A., E. L. Swenson, M. L. Wolfe, D. C. Choate, B. Bates, and K. A. Crandall. 2001. "Gene flow estimates in Utah's cougars imply management beyond Utah," *Animal Conservation* 4:257–264.

Soulé, M., and R. Noss. 1998. "Rewilding and biodiversity: Complementary goals for continental conservation," *Wild Earth* 8:18–28.

Sunquist, M., and F. Sunquist. 2002. *Wild Cats of the World.* University of Chicago Press, Chicago.

Sweanor, L. L, K. A. Logan, and M. G. Hornocker. 2000. "Cougar dispersal patterns, metapopulation dynamics, and conservation," *Conservation Biology* 14:798–808.

Teel, T. L., R. S. Krannich, and R. H. Schmidt. 2002. "Utah stakeholders' attitudes toward selected cougar and black bear management practices," *Wildlife Society Bulletin* 30:2–15.

Terborgh, J., J. A. Estes, P. C. Paquet, K. Ralls, D. Boyd-Heger, B. Miller, and R. Noss. 1999. "The role of top carnivores in regulating terrestrial ecosystems," in *Continental Conservation: Scientific Foundations of Regional Reserve Networks*, ed. M. E. Soulé and J. Terborgh, 60–103. Island Press, Washington, DC.

Terborgh, J., L. Lopez, P. Nuñez V., M. Rao, G. Shahabuddin, G. Orihuela, M. Riveros, R. Ascanio, G. H. Adler, T. D. Lambert, and L. Balbas. 2001. "Ecological meltdown in predator-free forest fragments," *Science* 294:1923–1926.

Torres, S., H. Keough, and D. Dawn. In prep. "Mountain lion management in western North America: A 100 year retrospective."

Torres, S. G., T. M. Mansfield, J. E. Foley, T. Lupo, and A. Brinkhaus. 1996. "Mountain lion and human activity in California: Testing speculations," *Wildlife Society Bulletin* 24:451–460.

U.S. Census Bureau. 2001. *Projections of the Total Populations of States: 1995 to 2055.* www.census.gov/population/projections/state/stpjpop.txt. Accessed September 15, 2003.

U.S. Fish and Wildlife Service (USFWS). 1987. *Florida Panther Recovery Plan.* Florida Panther Interagency Committee, Atlanta.

Walker, C. 2003. "Cougar reports on the rise in eastern U.S.," *National Geographic News*, March 7.

Washington Department of Fish and Wildlife. 2002. *Game Management Plan (Draft).* Wildlife Program, Washington Department of Fish and Wildlife, Olympia.

Wilcox, B. A., and D. D. Murphy. 1985. "Conservation strategy: The effects of fragmentation on extinction," *American Naturalist* 125:879–887.

Woodroffe, R. 2000. "Predators and people: Using human densities to interpret declines of large carnivores," *Animal Conservation* 3:165–173.

Young, S. P., and E. A. Goldman. 1946. *The Puma: Mysterious American Cat.* American Wildlife Institute, Washington, DC.

CHAPTER 12

Restoring the Gray Wolf to the Southern Rocky Mountains: Anatomy of a Campaign to Resolve a Conservation Issue

Michael K. Phillips, Rob Edward, and Tina Arapkiles

Habitat loss, habitat fragmentation, and persecution by humans over the last several decades have significantly reduced occupied historic range for most large carnivore species (Fuller 1995; Fuller and Kittredege 1996). An interest in recovering these species has spawned numerous restoration projects (Reading and Clark 1996) and a burgeoning number of articles and other scientific literature regarding the ecological importance of large carnivores (Estes et al. 1989; McLaren and Peterson 1994; Terborgh et al. 1999; Ripple and Larsen 2000; Berger et al. 2001; Ripple et al. 2001; Terborgh et al. 2001; Mech and Boitani, 2003; Mech and Peterson, 2003); the varied biological and sociopolitical aspects of large carnivore conservation (Clark et al. 1996; Keiter and Locke 1996; Kellert et al. 1996; Rasker and Hackman 1996; Weber and Rabinowitz 1996; Nie 2003); and the outcome of individual restoration projects (Reading and Clark 1996; Breitenmoser et al. 2001; Phillips et al. 2003; Smith et al. 2003).

In contrast, scant information exists on how to catalyze consideration of a large carnivore conservation or restoration project so that final decisions regarding implementation are based on relevant legal and scientific standards and are supported by most if not all of the affected parties. This chapter attempts to fill that gap by providing the details of a campaign to catalyze such consideration of restoring the gray wolf (*Canis lupus*) to the southern Rocky Mountain (SRM) region.

After centuries of persecution, by the late 1950s the number of gray wolves inhabiting the conterminous United States had reached an all-time low (Young and Goldman 1944; Young 1970; Brown 1983; Nowak 1983).

12. Restoring the Gray Wolf to the Southern Rocky Mountains

By then, less than 1% of the species' historic range was occupied by fewer than 1,000 wolves in the remote forests of northeastern Minnesota. Additionally, probably fewer than 20 wolves inhabited Isle Royale National Park, a 546 km^2 (210 mi^2) island in Lake Superior (Stenlund 1955; Mech 1966; Peterson 1977; Fuller et al. 1992; Thiel 1993). Passage of the Endangered Species Act (ESA) in 1973 provided protection for gray wolves and signaled a new era for conservation. Within 30 years, recovery programs relying on expansion of extant populations in Minnesota and Canada, along with reintroduction efforts in the northern Rocky Mountains and the southwestern United States, resulted in significant increases in wolf numbers and distribution. By March 2003, the species occupied slightly less than 5% of its historic range in the conterminous United States and included about 3,500 animals.

In response to the improved conservation status for gray wolves, in April 2003 the U.S. Fish and Wildlife Service (USFWS) released a reclassification rule indicating that recovery had essentially been completed in most of the conterminous United States and reduced federal protection for the species (USFWS 2003a). The reclassification rule divided the lower forty-eight states into three distinct population segments (DPS)—areas that support wolf populations, are somewhat separated from one another, are significant to the overall conservation of the species, and are considered separately under the ESA (Figure 12.1).

The USFWS's definition of recovery is important because further recovery activities will be difficult to implement once the species is removed from ESA listing. It is unlikely that state legislators, state game commissions, and corresponding state game agencies would initiate actions for recovering wolves once the species has been removed from the federal list of threatened and endangered species, as is shown by the following examples.

In 1995, the Montana Legislature passed Senate Bill 394, which amends Title 81 (Department of Livestock) sections by adding the wolf to the definition of predatory animal (81-7-101, MCA). Furthermore, it states: "The Department of Livestock shall conduct the destruction, extermination, and control of predatory animals capable of killing, destroying, maiming, or injuring domestic livestock or domestic poultry, and the protection and safeguarding of livestock and poultry in this state against depredations from these animals" (81-7-102, MCA). This section also states that the Department of Livestock shall "adopt rules applicable to

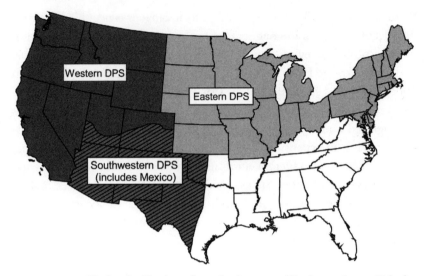

FIGURE 12.1 Final reclassification scheme for the gray wolf in the contiguous United States. USFWS 2003a.

predatory animal control which are necessary and proper for the systematic destruction of the predatory animals by hunting, trapping, and poisoning operations and payments of bounties." The effective date of this act is "whenever the gray wolf is removed from the list of threatened or endangered species by the appropriate agency of the United States government."

In September 1989, the Colorado Wildlife Commission adopted a resolution opposing reintroduction of the gray wolf. Colorado state law (#35-40-107 and 108) authorizes payment of a $2 bounty on wolves, and in October 1999 the Colorado General Assembly effected a bill that requires the state legislature to approve any reintroduction of a federally listed species (Colorado state law #33-2-105.5).

In March 2000, the New Mexico Wildlife Commission unanimously reaffirmed its opposition to reintroducing Mexican wolves.

In late 2001, legislators in Utah began considering passage of a state law that would require the USFWS to remove any wolf that dispersed to the state (E. E. Bangs, pers. comm.).

The final federal reclassification rule indicated that the recovery objectives for the eastern gray wolf DPS and the western gray wolf DPS were met in late winter 1999 and December 2002, respectively. Consequently, the USFWS determined that additional recovery activities for those two regions were unnecessary (USFWS 2003a, 2003b, 2003c). The final rule noted that recovery in the southwestern gray wolf DPS, on the other hand, was incomplete because wolf numbers were low and threats were high. Because of this the USFWS decided to classify the southwestern gray wolf DPS as endangered, except for animals in the experimental-nonessential population area that had been created to support the reintroduction of the Mexican wolf (*C. l. baileyi*) in the Blue Range Wolf Recovery Area (USFWS 1996, 1998). Experimental-nonessential populations are designated by the USFWS per section 10(j) of the ESA to minimize conflict between humans and members of an endangered species involved in a reintroduction project (Parker and Phillips 1991).

The determination on the status of the southwestern gray wolf DPS resulted from the USFWS's evaluation of the best scientific information available. This information indicated that any other determination was inappropriate. The ESA requires that a listed species be recovered across a significant portion of its historic range before it can be removed from the list of endangered and threatened wildlife (50 CFR 402.02, ESA section 4(a)(1), United States Ninth Circuit Court of Appeals 2001). Some nongovernmental conservation organizations (conservation NGOs) and scientists pointed out that wolf recovery in the southwestern United States would require that a large area in the region be made available for restoration efforts. The final configuration of the southwestern gray wolf DPS accomplished that (see Figure 12.1).

This chapter details the efforts of conservation NGOs to catalyze public consideration of wolf restoration in the southern Rocky Mountains region (SRM), a specific area within the southwestern gray wolf DPS. The campaign aims to foment a final decision that is based on the pertinent legal and scientific standards, and enjoys the support of most—if not all—of the affected parties. Because of the biological, logistical, fiscal, and sociopolitical complexity of wolf restoration, this chapter should be useful to others working to advance carnivore restoration and conservation projects that are mired in controversy, misunderstanding, and misinformation.

Southern Rocky Mountains and Gray Wolf Restoration

The effort to restore wolves to the SRM springs from the region's unique characteristics, including vast expanses of unroaded public land and burgeoning populations of wild ungulates. The SRM extends from south-central Wyoming through western Colorado into north-central New Mexico (Shinneman et al. 2000). The region includes 39,000 mi^2 of public land that supports sufficient prey to sustain a viable population of wolves, and currently represents a significant gap in the range of the species (Shinneman et al. 2000). The last wolf known from the SRM was shot in the San Juan Mountains in southern Colorado in 1945 (Bennett 1994).

The SRM region contains about one and a half times more public land than is available to wolves in the Greater Yellowstone Ecosystem (25,000 mi^2), and almost twice as much as is available in central Idaho (20,781 mi^2) (Figure 12.2). The region also contains about 6 times the amount of public land available to Mexican wolves in the Blue Range Wolf Recovery Area (BRWRA) in southeastern Arizona and southwestern New Mexico (6,854 mi^2). Lastly, the SRM region includes 1.7 to 25 times more public land than other sites considered by some to be appropriate for wolf restoration (Ferris et al. 1999).

Government agencies and some private landowners presently manage extensive tracts of land in the SRM in a fashion that could facilitate wolf recovery. The region includes significant roadless areas covering about 14,000 mi^2 and about 7,000 mi^2 of legally designated or de facto wilderness (Shinneman et al. 2000), which equals 70% of the wilderness available to wolves in the Yellowstone area. It is also comparable to the amount of wilderness available to wolves in central Idaho, and is about four times the amount of wilderness available to Mexican wolves in the BRWRA. Not surprisingly, a congressionally mandated study concluded that the Colorado portion of the SRM could support over 1,000 wolves (Bennett 1994). Additionally, a 1994 public opinion poll revealed that about 71% of registered voters in Colorado supported the restoration of gray wolves to the state (Manfredo et al. 1994; Pate et al. 1996).

Some believe that wolf recovery in the SRM could be especially significant from a continental perspective. Because the ecoregion is nearly equidistant from the northern Rocky Mountains and the BRWRA, it is possible that a SRM population would contribute significantly to the establishment and maintenance of a spatially segregated population of wolves that extended from the Arctic to Mexico. On the significance of wolf restoration to the SRM, noted wolf authority Dr. L. David Mech (2000) wrote: "Ultimately, then, this restoration could connect the entire North American wolf population from Minnesota, Wisconsin, and Michigan through Canada

12. *Restoring the Gray Wolf to the Southern Rocky Mountains* 245

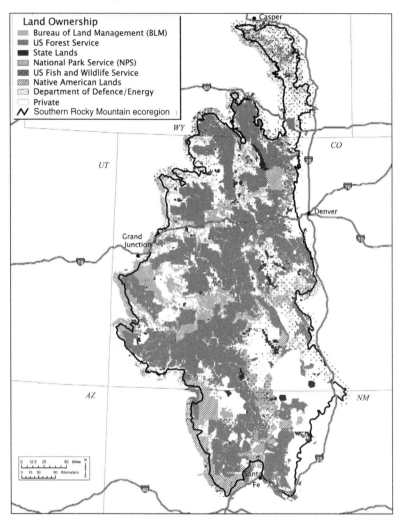

FIGURE 12.2 Land ownership patterns in the southern Rocky Mountain ecoregion. Shinneman et al. 2000.

and Alaska, down the Rocky Mountains into Mexico. It would be difficult to overestimate the biological and conservation value of this achievement."

Given the strong public support for wolf restoration in the SRM, and the maturation of existing wolf reintroduction projects elsewhere in the lower forty-eight states, a group of experts met at Vermejo Park Ranch in northern New Mexico during December 1997 to consider the issue of restoring wolves to the SRM. That meeting gave rise to recognition that three steps were requisite to resolving the issue: (1) development of coalitions that would focus on advocacy, education, and research; (2) development and dissemination of the best available science on the matter; and (3) development of an outreach program to engage affected parties in serious and comprehensive discussions about restoring wolves to the SRM.

Step 1: Development of Coalitions

Wolf restoration is a controversial issue and one that federal and state agencies and elected officials are often reluctant to consider. For example, even though well-respected biologists were arguing for restoring the wolf to Yellowstone Park as early as the 1940s (Leopold 1944) and some within the Department of Interior openly discussed the issue during the 1970s (Weaver 1978), it was not until 1994 that the issue was finally resolved (USFWS 1994). Because of this institutional and political reluctance, it was concluded that serious consideration of the issue could only be advanced through the formation of a coalition that strongly advocated wolf restoration—ultimately leading to the formation of the Southern Rockies Wolf Restoration Project in February 2000.

Coalitions to Promote Restoration

The Southern Rockies Wolf Restoration Project is a coalition of regional and national conservation organizations dedicated to restoring wolves to their full ecological role throughout the SRM. Member groups include Sinapu, Sierra Club (national), the Wildlands Project, Defenders of Wildlife, National Wildlife Federation, the Center for Biological Diversity, the New Mexico Wildlife Federation, and the New Mexico Wilderness Alliance. Supporting groups include several other regional and national conservation organizations.

The project's efforts are rooted in the principles of conservation biology and conflict resolution. Much of its work must satisfy the scrutiny of their scientific advisory team, a group of well-respected biologists with extensive experience with wolf restoration. Working from the best available science, the project engages in discussions with all who are interested in wolf restoration—from ranchers, to hunters, to conservation and wildlife organizations, to policymakers—based on the premise that open-minded communication and careful consideration of credible scientific and policy information will ultimately result in wolf restoration in the SRM. Adherence to that premise has, thus far, spawned several interesting endeavors.

Since inception, the project has developed a population and habitat viability assessment for wolves in the region (Phillips et al. 2000); cosponsored a public opinion poll (Meadow 2001; Meadow et al., in press); cosponsored the development of a feasibility study of the social aspects of wolf restoration (Phillips et al., in prep.); and successfully challenged the USFWS to consider wolf recovery in the SRM region in light of relevant scientific and legal standards rather than standards applicable only to wolf recovery in the northern Rocky Mountains.

This final point is extremely important. In response to the improved conservation status of gray wolves in the northern Rocky Mountains and the Great Lakes region by the late 1990s, the USFWS began developing a new national vision for wolf recovery. This vision first appeared in July 2000 as a proposed reclassification rule (USFWS 2000). The proposal indicated that the SRM was best considered as part of the western gray wolf DPS. The proposal further indicated that as soon as recovery objectives developed for the northern Rocky Mountains were realized, then the USFWS would delist the western gray wolf DPS, thus precluding any additional federal restoration efforts anywhere in the DPS, including the SRM, and squelching any further effort to restore wolves to the SRM.

For two years following release of the proposed rule, the Southern Rockies Wolf Restoration Project worked diligently to convince the USFWS that the proposed rule was inconsistent with relevant scientific and legal standards for recovery. The project successfully advanced its message by submitting comments on the proposal; by educating its membership about the proposal, thus generating comments from citizens from around the country and overwhelming participation at regional hearings; by writing letters to the editor in regionwide newspapers; by encouraging the interest of elected officials; by effecting strategic and frequent communication with the USFWS; and by alerting state game agencies to the deficiencies of the proposed rule. This effort spurred high-ranking officials in Idaho, Montana, Wyoming, Colorado, and New Mexico to send a rare joint letter to the director of the USFWS requesting a recision of the proposed rule. Letters from members of the U.S. Congress to the USFWS also resulted from the coordinated efforts of the project.

While the USFWS did not rescind the proposal, the project was partially successful since the final reclassification rule (USFWS 2003a) shifted the southern two-thirds of the SRM into the southwestern gray wolf DPS (see Figure 12.1), thus maintaining endangered classification for much of the region. This decision represents tacit recognition that wolf recovery in the southwestern United States (including the southern half of the SRM) is far from complete, setting the stage for comprehensive and science-based recovery planning and implementation. Although somewhat successful in the political advocacy arena, the project's efforts require ongoing complementary strides within the realms of constituent education and biological research/monitoring.

Coalitions Dedicated to Education and Outreach

The second type of coalition we formed was dedicated to education and outreach and was named the Wolf Forum for the Southern Rockies. Disseminating relevant and credible information to affected parties is an important component of any large carnivore restoration project (Reading and Clark 1996), replacing exaggeration and misinformation with facts and science-based estimations of the results of possible actions. Recognition of this catalyzed the formation of the Wolf Forum for the Southern Rockies.

The forum aims to disseminate scientific information and position statements from diverse organizations that either oppose or support wolf restoration in the ecoregion. In order to remain impartial, the forum will not take a position on wolf restoration. Member groups include the following public educational zoological institutions: the Denver Zoo, the Cheyenne Mountain Zoo, the Albuquerque Biological Park, the Pueblo Zoo, and the International Wolf Center.

The forum operates two complementary activities rooted in the premise that "good information provided to good people leads to good decisions." First, it disseminates diverse viewpoints on wolf restoration. Any recognized professional organization that has a stake in the issue is invited to submit to the forum its position statement regarding restoring wolves to the SRM. If the position statement is determined by the forum's steering committee to be respectful and professional in tone, then it will be posted on the forum's website. Such postings will ensure that interested parties can review various perspectives on the issue.

Second, the forum disseminates credible scientific information. The forum will accept scientific publications for review by their science advisory team. If the advisory team determines that a submission is based primarily on peer-reviewed or peer-edited scientific literature and presents a professional and balanced analysis of data, then the team will recommend endorsement of the publication by the forum. The forum will post endorsed publications on the website and actively distribute the publication to the public at each of the member facilities and at off-site outreach events.

Because of the forum's emphasis on information dissemination, its greatest asset is the member groups' relationships with 3 million patrons and members. Consequently, the future addition of other zoological/ecological educational institutions will expand the effectiveness of the forum.

Coalitions Dedicated to Research

The third type of coalition we developed was dedicated to research and was called the Southern Rockies Wolf Monitoring and Research Team. Research and monitoring are important components of any responsible restoration project (IUCN 1998) and scientists have documented previous wolf restoration projects, yielding volumes of new information (Phillips and Smith 1996; Bangs et al. 1998; Fritts et al. 2001; Paquet et al. 2001; Phillips et al. 2003). Clearly, if a wolf restoration project in the SRM moved forward, it would require high-quality research and monitoring. This recognition catalyzed the formation of the Southern Rockies Wolf Monitoring and Research Team.

The team consists of biologists from the Turner Endangered Species Fund, the U.S. Geological Survey Biological Resources Division, Michigan Technological University, the Wildlife Conservation Society, the Wildlands Project, and the Denver Zoological Foundation. The team has committed to providing fiscal, logistical, and intellectual support—along with at least 700,000 acres of livestock-free, nearly contiguous, high-quality habitat where manipulative experiments and field studies could be performed. Notably, this public-private partnership could do much to ameliorate fiscal criticism by wolf restoration opponents, because it would maximize the amount of private funding that is used to offset the cost of research and monitoring.

Step 2: Development and Dissemination of the Best Available Science

Section 4(b) of the ESA and related USFWS policies clearly emphasize that decisions regarding listed species must be based on the best available scientific and commercial data. Before 1999, only Bennett (1994) had attempted to estimate the biological potential of the SRM to support wolves. While his work indicated that western Colorado could support a population of 1,000 wolves, the analysis was limited in scope (only national forest lands in western Colorado were evaluated) and generalized across the landscape. Improved habitat suitability estimates for the entire SRM under current and future conditions were needed based on much improved analytical techniques (Carroll et al. 2001a, 2001b). Further, there was a need for a new public opinion poll to determine if majority support for wolf restoration had changed since the mid-1990s. Lastly, there was need for a research effort to combine the myriad biological, social, economic, and political aspects of the issue in a comprehensive assessment.

Localized Study of Habitat Suitability

In 1999, researchers initiated an assessment of the ecological and socioeconomic factors that might influence wolf restoration over nearly 3,100 mi^2 in the southern portion of the SRM. The study included 418 mi^2 of the Carson National Forest, the privately owned Vermejo Park Ranch (covering 910 mi^2), and six additional parcels (covering 734 mi^2) of private land that are managed for conservation purposes. Results indicated that this relatively small area could easily support 100 wolves or more owing to adequate prey populations, favorable patterns of land ownership and management, and potentially few conflicts with livestock (Southern Rockies Ecosystem Project 2000).

Wolf Population and Habitat Viability Assessment

In August 2000, several interested groups organized a population and habitat viability assessment (PHVA) for gray wolves in the SRM, with the respected Conservation Breeding Specialist Group (CBSG) serving as an expert and neutral facilitator and organizer. CBSG is a member of the Species Survival Commission of the IUCN—World Conservation Union, and for more than a decade has been developing, testing, and applying science-based tools and processes to assist with management of imperiled species and imperiled biomes. By bringing together scientists, landowners, wildlife agency personnel, conservationists, ranchers, hunters, and interested individuals, the PHVA catalyzed a broad public discussion about restoring wolves to the SRM (Phillips et al. 2000).

Notably, participation in the PHVA by organizations and individuals did not imply support for wolf restoration; rather, it served as an opportunity to share views and expertise on biological and sociological issues relevant to wolf restoration. The workshop created an opportunity for participants to share ideas, served as a forum to discuss the implications of wolf restoration to the region, and used modeling to identify potential habitat for wolves by illuminating factors such as prey and road density. The final report indicated that the biological suitability of the SRM for gray wolves was very high and that the issue was characterized by complex and onerous social and political challenges (Phillips et al. 2000).

Regional Study of Habitat Suitability

In 2001, researchers who had participated in the PHVA initiated a study using static and dynamic spatial models to evaluate suitability of the SRM for reintroduction of gray wolves (Carroll et al. 2003). In particular, the study questioned whether reintroduction would advance restoration by increasing the species' distribution beyond what might be expected through natural range expansion by wolves inhabiting Montana, Wyoming, or Idaho (i.e., the northern Rocky Mountains).

Carroll et al. (2003) used multiple logistic regression to develop a resource-selection function relating wolf distribution in the Greater Yellowstone region with regional-scale habitat variables for the SRM. Results indicated that areas of the SRM with resource-selection-function values similar to those of currently inhabited areas in Yellowstone could potentially support more than 1,000 wolves. They also employed a spatially explicit dynamic population model (i.e., PATCH model) to predict wolf distribution and viability at several potential reintroduction sites within the SRM under current landscape conditions and two contrasting predictions of future landscape conditions. The PATCH model predicted a wolf population of greater than 1,000 animals but indicated that development trends over the next 25 years may result in the loss of one of four potential regional subpopulations and increased isolation of the remaining three (Carroll et al. 2003). The results also indicated that, owing to the low level of connectivity between the Yellowstone area and the SRM, there was virtually no chance for a wolf population to become established in the SRM via recolonization by wolves dispersing from the Yellowstone area. The findings by Carroll ct al. (2003) corroborated the findings of previous studies (Bennett 1994; Martin et al. 1999) that indicated that the SRM could support a self-sustaining population of wolves (Table 12.1).

To further clarify the regional value of the SRM as a gray wolf restoration site, in 2002 researchers initiated a comprehensive assessment of potential habitat, landscape-level threats, and population viability for gray

TABLE 12.1
Results of various research projects to estimate the capacity of the southern Rocky Mountains to support wolves

Model and Scenario	Number of Packs	Mean Pack Size (adults)	Mean Pop. Size (including pups)
PATC—Landscape scenario A[a] (Carroll et al. 2003)	141.5	5.10	1,337
PATCH—Landscape scenario B[b] (Carroll et al. 2003)	72.7	4.93	664
PATCH—Landscape scenario C[c] (Carroll et al. 2003)	48.0	4.85	431
Resource selection function (RSF) (Carroll et al. 2003)	N/A	N/A	1,305
Bennett (1994) (Colorado only)	N/A	N/A	500–1,000
Martin et al. (1999) (Colorado only)	67	~5.4	670

[a]Landscape scenario A = current conditions in the southern Rocky Mountains.
[b]Landscape scenario B = human population as projected for 2025 (U.S. Census Bureau, unpublished data) with increased road development (Theobald et al. 1996) on private lands only.
[c]Landscape scenario C = human population as projected for 2025 (U.S. Census Bureau, unpublished data) with increased road development (Theobald et al. 1996) on private and unprotected public lands.

wolves (and jaguars, *Panthera onca*) across the southwestern United States and northern Mexico (Carroll et al., in prep.). The study area includes southeastern Utah, all of Arizona, most of New Mexico, southwestern Texas, and several states in Mexico, including Chihuahua, Coahuila, Nuevo Leon, Durango, and portions of Tamaulipas and Zacatecas. This area encompasses the majority of the estimated historic distribution of the Mexican wolf, excluding the southern extreme of its range, and it encompasses nearly all of the Southwestern DPS not considered by Carroll et al. (2003).

Results from this work and Carroll et al. (2003) will serve as the basis for a regional-scale conservation strategy that: (1) addresses specific threats (landscape change and development) to wolf (and jaguar) population recovery in the southwestern DPS and northern Mexico; (2) prioritizes areas for wolf (and jaguar) restoration; and (3) compares the efficacy of alternative restoration strategies.

Comprehensive Wolf Restoration Feasibility Report

In 1988, the U.S. Senate–House Interior Appropriations Committee posed four questions to the U.S. Fish and Wildlife Service and U.S. National Park Service about restoring wolves to Yellowstone National Park and environs. The questions resulted in the *Wolves for Yellowstone?* studies, which presented more than 1,200 pages of evaluation that greatly advanced resolution of the issue of restoring wolves to Yellowstone (Yellowstone National Park et al. 1990; Varley and Brewster 1992).

In 2002, taking a cue from the outcome of the *Wolves for Yellowstone?* studies, researchers began developing the first comprehensive report on the biological, economic, and sociopolitical aspects of restoring wolves to the SRM. The authors of the report, "Suitability of the Southern Rockies for Wolf Restoration: An Ecological and Social Assessment" (Phillips et al., in prep.), aim to use a science-based approach to assemble all information relevant to the issue. The report will not advocate for or against restoration but will instead simply present the facts of the issue. As such, the report will facilitate an evolution of the discussion of the issue from the current misinformed and sometimes exaggerated debate to a discourse characterized by well-reasoned positions based on relevant legal, biological, and social standards.

Public Opinion Polls

Public attitudes about gray wolves have greatly affected the fortunes of the species. For example, the belief that wolves represented a significant impediment to the country's manifest destiny to tame the wilderness fueled a federal extermination policy that persisted for more than a century (Phillips et al., in press). By the early 1970s, however, attitudes toward nature in general and wolves specifically had changed. By this time, a strong pro-wolf constituency arose from a public perception that the species had an inherent right to exist and was important for maintaining ecosystem health. Given the connection between public attitudes and the viability of wolf populations, the necessity to accurately gauge public attitudes about gray wolves in the SRM is clear.

In 1994, a public opinion poll revealed that about 71% of registered voters in Colorado supported the restoration of gray wolves to Colorado (Manfredo et al. 1994; Pate et al. 1996). By late 2000, however, Manfredo et al. (1994) could only be considered a data point, rather than a gauge of current public opinion on the subject. Further, the survey focused exclusively on Colorado, leaving the rest of the SRM—and the Southwestern DPS—as significant gaps. Consequently, Decision Research conducted new public opinion research concerning the restoration of wolves to the SRM by polling 500 registered voters in Colorado and 400 in each New Mexico and Arizona. The survey revealed strong support for wolf restoration (Meadow 2001; Meadow et al., in press). In each of the states, a majority of voters favored restoring wolves into national forest and wilderness areas. Support was widespread, nonpartisan, and exceeded 60% for seven of eight demographic groups, including males and females, Republicans and Democrats, hunting households and nonhunters. At the conclusion of the survey, after voters had heard arguments both for and against restoration, support for wolf restoration increased to 70% overall, thus indicating the importance of public education to increase support for wolf restoration.

Upon completion of the poll, the sponsoring organizations launched a comprehensive public relations effort to release the findings. This effort included notification to local, regional, and national media and direct correspondence with high-ranking state and federal officials. Additionally, the sponsoring organizations arranged two meetings (in Denver, Colorado, and Albuquerque, New Mexico) for state and federal officials and interested citizens to meet directly with Decision Research, and to hear Decision Research explain their scientific methods for collecting, analyzing, and interpreting data.

Study of the Economics of Wolf Restoration

As wolf recovery has proceeded throughout the United States, some groups have voiced concern about the species' potential to negatively impact local and regional economies. Specifically, the livestock and hunting industries have claimed that wolf recovery is bad for business. Research, however, does not support these claims, but rather demonstrates that a wolf population and wolf recovery and restoration activities can generate positive economic benefits. For example, Schaller (1996) indicated that the International Wolf Center in Ely, Minnesota, generates about $3 million in annual economic activity and as many as 66 new jobs in the region. The center is based in Ely because of the presence of a wolf population in northern Minnesota. A study done by Cornell University asserts that red wolf restoration at the Alligator River National Wildlife Refuge generated an annual average regional economic benefit of about $37.5 million because of increased tourism spawned by the project (Rosen 1997). Before the reintroduction of wolves to Yellowstone Park, economists from the University of Montana estimated the project's net economic benefits at up to $23 million annually from increased visitor expenditure (USFWS 1994).

Since wolves were released in Yellowstone in March 1995, researchers have collected original data that will more precisely gauge the project's economic effects. Recently, the Yellowstone Park Foundation initiated a study to accomplish this task (Yellowstone Park Foundation 2003). Completion of the study will have immediate bearing on the issue of restoring gray wolves to the SRM by properly accounting for the costs and benefits of the species' presence in a portion of the northern Rocky Mountains.

Step 3: Development of Education and Outreach

Since 1999, those working toward a full discussion of wolf restoration in the SRM have embraced the idea that such a discussion will arise only from a broad-based education and outreach program aimed at interested and affected parties. Proponents of restoration have repeatedly discussed the issue with high-ranking officials with the Departments of Interior and Agriculture; members of the U.S. Congress; and directors and senior staff from the U.S. Fish and Wildlife Service, the U.S. Forest Service, and state game agencies. Thus far, proponents of restoration have given more than 50 presentations at professional conferences and invited lectures hosted by federal and state agencies, conservation NGOs, and universities. Proponents have discussed wolf restoration with several owners of large tracts of land in northern New Mexico and southern Colorado. Lastly, proponents have published articles in membership magazines and newsletters and granted numerous interviews with regional and national media about restoring wolves to the SRM.

With much new information regarding wolf restoration in general (Phillips and Smith 1996; Bangs et al. 1998; Brown and Parsons 2001; Smith et al. 2003)—and restoration to the SRM in particular (Phillips et al. 2000; Carroll et al. 2003; Meadow et al., in press)—now appearing in peer-reviewed journals and other technical publications, those advocating a discussion of restoration intend to integrate such information in a sophisticated fashion into future outreach efforts. As mentioned earlier, the technical report titled "Suitability of the Southern Rockies for Wolf Restoration: An Ecological and Social Assessment" (Phillips et al., in prep.) will present several hundred pages of technical information. The report's length and technical nature are not designed to have the general public as the primary audience. Consequently, an effort is planned to integrate the report's highlights into a popular version called "The Truth About Wolves." Watershed Media, a nonprofit organization that produces communication projects to influence the transition to a more sustainable society, is working with the Southern Rockies Wolf Restoration Project and the Wolf Forum for the Southern Rockies to create a visually dynamic book that will serve as the centerpiece for future outreach efforts. In addition to the book and technical report, the Truth About Wolves campaign will include concerted outreach to the local, regional, and national media; three public symposia to be held in Colorado, New Mexico, and Washington, D.C.; a series of smaller public outreach events; and free web versions of "The Truth About Wolves" and related educational material. The Truth About Wolves campaign will be designed to project a clear sense of objectivity and scientific fact, rather than heated advocacy. The campaign is scheduled to begin in 2005.

These outreach efforts aim to convince federal and state game agencies—and all other interested and affected parties—that a comprehensive discussion about restoring wolves to the SRM is appropriate. Four factors support this claim: (1) the region's high biological suitability; (2) the broad and persistent bipartisan public support for the idea; (3) the USFWS's decision to retain an endangered status for the species throughout the southern half of the SRM; and (4) the strong relevance of the scientific and legal standards that guide endangered species recovery. A comprehensive discussion should result in the identification and proper consideration of the salient challenges of restoring gray wolves to the SRM. Such an outcome should greatly facilitate development of a final resolution to the issue that is acceptable to most, if not all, stakeholders.

Conclusion

Restoring gray wolves is a highly emotional and politicized issue, made more so by the biological success of the species in Great Lakes region and the northern Rocky Mountains. Even though the species still occupies less than 5% of its historic range, wolf population numbers today exceed those of the early 1900s. While some claim that recovery requires greater occupation of historic range, others assert that additional restoration efforts are superfluous for a species that is represented by several thousand individuals. Regardless of one's position on the issue, the southwestern gray wolf DPS, which includes over 50% of the SRM, retains an endangered classification, thus requiring the USFWS to develop and implement a recovery plan to repatriate the species to a significant portion of its historic range there. Consequently, the issue of restoring wolves to the SRM remains relevant and unresolved. A comprehensive campaign is now underway to use relevant scientific and legal standards and concerted outreach and education efforts to achieve a final resolution to the issue that will be supported by a majority of the affected parties. In the end, the success of this campaign will not be measured by whether wolves are restored to the SRM, but by how clearly affected parties understand the facts at the core of the debate.

Much future work will be required to press the importance of an objective, science-based approach to resolving the issue of restoring wolves to the SRM. The coalitions that have formed around the issue should immediately and obviously support efforts by the USFWS to develop a recovery plan, which is the next most important step in reaching a resolution.

The challenges related to developing a recovery plan notwithstanding, implementation will require a much greater effort. After a recovery plan is approved, consideration should be given to forming a citizen management committee that would assume differential responsibility for this most difficult task. This approach was very useful in advancing a broadly supported plan for restoring grizzly bears to central Idaho (USFWS 2000; Roy et al. 2001). Both the Southern Rockies Wolf Restoration Project and the Wolf Forum for the Southern Rockies could do much to advance the idea of a citizen management committee for wolves.

Many who work on predator restoration and conservation initiatives have concluded that facts and good science are necessary for advancing their work. Indeed, this book is testimony to that fact. However, often it is neither the scientists nor conservationists who decide which efforts will be funded or what research will be used as a reference for policy decisions about predators. In democratic societies, politicians are usually charged with making such decisions. Like all of us, politicians are often motivated by expediency, experience, and self-interest rather than by knowledge, especially when the facts to do not support their values and beliefs (Petty et al. 1997). While knowledge helps to determine an individual's values and attitudes, which significantly impact decision making, its importance if often overestimated, especially among people who value knowledge greatly, such as scientists, conservationists, and ecological restorationists (Reading 1993; Kellert et al. 1996). Consequently, politicians are somewhat innately inclined, and most are lobbied frequently, to discount the best science available, especially when science-based answers about predator restoration and conservation are difficult to understand, do not support their values and beliefs, and counter their vested interests.

Despite these limitations, the political process is our best crucible for grinding the requisite public debate that will foment the social changes necessary for effective restoration and conservation of predators. The past and future activities described in this chapter should contribute much to that onerous but absolutely essential process.

Literature Cited

Bangs, E. E., S. H. Fritts, J. A. Fontaine, D. W. Smith, K. M. Murphy, C. M. Mack, and C. C. Niemeyer. 1998. "Status of gray wolf restoration in Montana, Idaho, and Wyoming," *Wildlife Society Bulletin* 26:785-798.

Bennett, L. E. 1994. *Colorado Gray Wolf Recovery: A Biological Feasibility Study*. Final Report. U.S. Fish and Wildlife Service and University of Wyoming Fish and Wildlife Cooperative research unit, Laramie, WY.

Berger, J., P. B. Stacey, M. L. Johnson, and L. Bellis. 2001. "A mammalian predator-prey imbalance: Grizzly bear and wolf extinction affects avian neotropical migrants," *Ecological Applications* 11:947-960.

Breitenmoser, U., C. Breitenmoser-Wursten, L. N. Carbyn, and S. M. Funk. 2001. "Assessment of carnivore reintroductions," in *Carnivore Conservation*, ed. J. L. Gittleman, S. M. Funk, D. W. MacDonald, and R. K. Wayne, 241-281. Cambridge University Press, Cambridge.

Brown, D. E. 1983. *The Wolf in the Southwest: The Making of an Endangered Species*. University of Arizona Press, Tucson.

Brown, W. M., and D. R. Parsons. 2001. "Restoring the Mexican gray wolf to the desert southwest," in *Large Mammal Restoration: Ecological and Sociological Challenges in the 21st Century*, ed. D. S. Maehr, R. F. Noss, and J. L. Larkin, 169-187. Island Press, Washington, DC.

Carroll, C., R. F. Noss, and C. A. Lopez Gonzalez. In prep. "Spatial analysis of restoration potential and population viability of the Mexican wolf (*Canis lupus baileyi*) and jaguar (*Panthera onca*) in the southwestern United States and northern Mexico," *Conservation Biology*.

Carroll, C, R. F. Noss, and P. C. Paquet. 2001a. "Carnivores as focal species for conservation planning in the Rocky Mountain region," *Ecological Applications* 11:961-980.

Carroll, C., R. F. Noss, N. H. Schumaker, and P. C. Paquet. 2001b. "Is the return of the wolf, wolverine, and grizzly bear to Oregon and California biologically feasible?" in *Large Mammal Restoration: Ecological and Sociological Challenges in the 21st Century*, ed. D. S. Maehr, R. F. Noss, and J. L. Larkin, 25-46. Island Press, Washington, DC.

Carroll, C, M. K. Phillips, N. H. Schumaker, and D. W. Smith. 2003. "Impacts of landscape change on wolf restoration success: Planning a reintroduction program using static and dynamic models," *Conservation Biology* 17:536-548.

Clark, T. W., A. P. Curlee, and R. P. Reading. 1996. "Crafting effective solutions to the large carnivore conservation problem," *Conservation Biology* 10:940-948.

Estes, J. A., D. O. Duggins, and G. B. Rathbun. 1989. "The ecology of extinctions in kelp forest communities," *Conservation Biology* 3:252-264.

Ferris, R. M., M. Shaffer, N. Fascione, H. Pellet, and M. Senatore. 1999. *Places for Wolves: A Blueprint for Restoration and Long-Term Recovery in the Lower 48 States*. Defenders of Wildlife, Washington, DC.

Fritts, S. H., C. M. Mack, D. W. Smith, K. M. Murphy, M. K. Phillips, M.D. Jimenez, E. E. Bangs, J. A. Fontaine, C. C. Niemeyer, W. G. Brewster, and T. J. Kaminski. 2001. "Outcomes of hard and soft releases of wolves in central Idaho and the Greater Yellowstone Area," in *Large*

Mammal Restoration: Ecological and Sociological Challenges in the 21st Century, ed. D. S. Maehr, R. F. Noss, and J. L. Larkin, 125–147. Island Press, Washington, DC.

Fuller, T. K. 1995. "An international review of large carnivore conservation status," in *Integrating People and Wildlife for a Sustainable Future,* ed. J. A. Bissonette and P. R. Krausman, 410–412. The Wildlife Society, Bethesda, MD.

Fuller, T. K., W. E. Berg, G. L. Radde, M. S. Lenarz, and G. B. Joselyn. 1992. "A history and current estimate of wolf distribution and numbers in north central Minnesota," *Wildlife Society Bulletin* 20:42–54.

Fuller, T. K., and D. B. Kittredge Jr. 1996. "Conservation of large forest carnivores," in *Conservation of Faunal Diversity in Forested Landscapes,* ed. R. M. DeGraaf and R. I. Miller, 137–165. Chapman and Hall, New York.

IUCN/SSC. 1998. *Guidelines for Reintroductions.* IUCN Species Specialist Groups, IUCN, Gland, Switzerland, and Cambridge, UK.

Keiter, R. B, and H. Locke. 1996. "Law and large carnivore conservation in the Rocky Mountains of the U.S. and Canada," *Conservation Biology* 10:1003–1012.

Kellert, S. R., M. Black, C. R. Rush, and A. J. Bath. 1996. "Human culture and the conservation of large carnivores," *Conservation Biology* 10:977–990.

Leopold, A. 1944. "Review of the *Wolves of North America,* by S. P. Young and E. A. Goldman," *Journal of Forestry* 42:928–929.

Manfredo, M. J., A. D. Bright, J. Pate, and G. Tischbein. 1994. *Colorado Residents' Attitudes and Perceptions Toward Reintroduction of the Gray Wolf* (Canis lupus) *into Colorado.* Project Report No. 21, Human Dimensions in Natural Resources Unit, Colorado State University, Fort Collins.

Martin, B., R. Edward, and A. Jones. 1999. "Mapping a future for wolves in the southern Rockies," *Southern Rockies Wolf Tracks* 7:1–12.

McLaren, B. E., and R. O. Peterson. 1994. "Wolves, moose, and tree rings on Isle Royale," *Science* 266:1555–1558.

Meadow, B. 2001. *Southern Rockies Wildlife and Wilderness Survey Report.* Decision Research, Washington, DC.

Meadow, B., R. P. Reading, M. K. Phillips, M. Mehringer, and B. J. Miller. In press. "The influence of persuasive arguments on public attitudes toward a proposed wolf restoration in the southern Rockies," *Wildlife Society Bulletin.*

Mech, L. D. 1966. The Wolves of Isle Royale. Fauna Series No. 7. U.S. National Park Service, Washington, DC.

———. 2000. Comments on proposed wolf reclassification rule. Letter to U.S. Fish and Wildlife Service, Fort Snelling, MN, November 9, 2000.

Mech, L. D., and L. Boitani. 2003. "Ecosystem effects of wolves," in Wolves: Behavior, Ecology, and Conservation ed. L. D. Mech and L. Boitani, pp 158–160. University of Chicago Press, Chicago.

Mech, L. D., and R. O. Peterson. 2003. "Wolf-prey relations," in Wolves: Behavior, Ecology, and Conservation ed. L. D. Mech and L. Boitani, pp 131–160. University of Chicago Press, Chicago.

Nie, M. A. 2003. *Beyond Wolves: The Politics of Wolf Recovery and Management.* University of Minnesota Press, Minneapolis.

Nowak, R. M. 1983. "A perspective on the taxonomy of wolves in North America," in *Wolves in Canada and Alaska: Their Status, Biology, and Management,* ed. L. N. Carbyn, 10–19. Canadian Wildlife Service Report Series No. 45, Edmonton, AB.

Paquet, P. C., J. Vucetich, M. K. Phillips, and L. Vucetich. 2001. *Mexican Wolf Recovery: Three-year Program Review and Assessment.* Report prepared by the IUCN-SSC Conservation

Breeding Specialist Group, Apple Valley, MN, for the U.S. Fish and Wildlife Service, Albuquerque, NM.

Parker, W. T., and M. K. Phillips. 1991. "Application of the experimental population designation to the recovery of endangered red wolves," *Wildlife Society Bulletin* 19:73–79.

Pate, J., M. J. Manfredo, A. D. Bight, and G. Tischbein. 1996. "Coloradoans' attitudes toward reintroducing the gray wolf into Colorado," *Wildlife Society Bulletin* 24:421–428.

Peterson, R. O. 1977. *Wolf Ecology and Prey Relationships on Isle Royale*. National Park Service Scientific Monograph Series No. 11. U.S. Government Printing Office, Washington, DC.

Petty, R. E., D. T. Wegener, and R. L. Fabrigar. 1997. "Attitudes and attitude change," *Annual Review of Psychology* 48:609–674.

Phillips, M. K., E. E. Bangs, L. D. Mech, B. T. Kelly, and B. Fazio. In press. "Extermination and recovery of red wolf and gray wolf in the contiguous United States," in *The Biology and Conservation of Wild Canids*, ed. D. W. MacDonald and C. Sillero. Oxford University Press, Oxford.

Phillips, M. K., N. Fascione, P. Miller, and O. Byers. 2000. *Wolves in the Southern Rockies: A Population and Habitat Viability Assessment*. Final Report. IUCN/SSC Conservation Breeding Specialist Group, Apple Valley, MN.

Phillips, M. K., V. G. Henry, and B. T. Kelly. 2003. "Restoration of the red wolf," in *Wolves: Behavior, Ecology, and Conservation*, ed. L. D. Mech and L. Boitani, 272–288. University of Chicago Press, Chicago.

Phillips, M. K., B. J. Miller, R. Reading, R. Edward, and B. MacPherson. In prep. *Suitability of the Southern Rockies for Wolf Restoration: An Ecological and Social Assessment*. Turner Endangered Species Fund, Denver Zoological Foundation, Southern Rockies Wolf Restoration Project.

Phillips, M. K., and D. W. Smith. 1996. *The Wolves of Yellowstone*. Voyageur Press, Stillwater, MN.

Rasker, R., and A. Hackman. 1996. "Economic development and the conservation of large carnivores," *Conservation Biology* 991–1002.

Reading, R. P. 1993. "Toward an endangered species reintroduction paradigm: A case study of the black-footed ferret." Ph.D. dissertation, Yale University, New Haven, CT.

Reading, R. P., and T. W. Clark. 1996. "Carnivore reintroductions: An interdisciplinary examination," in *Carnivore Behavior, Ecology, and Evolution*, vol. 2, ed. J. L. Gittleman, 296–336. Cornell University, Ithaca, NY.

Ripple, W. J., and E. J. Larsen. 2000. "Historic aspen recruitment, elk, and wolves in northern Yellowstone National Park, USA," *Biological Conservation* 95:361–370.

Ripple, W. J., E. J. Larsen, R. A. Renkin, and D. W. Smith. 2001. "Trophic cascade among wolves, elk, and aspen on Yellowstone National Park's northern range," *Biological Conservation* 102:227–234.

Rosen, W. 1997. *Red Wolf Recovery in Northeastern North Carolina and the Great Smoky Mountains National Park: Public Attitudes and Economic Impacts*. College of Human Ecology, Cornell University, Ithaca, NY.

Roy, J., C. Servheen, W. Kasworm, and J. Waller. 2001. "Restoration of grizzly bears to the Bitterroot wilderness: The EIS approach," in *Large Mammal Restoration: Ecological and Sociological Challenges in the 21st Century*, ed. D. S. Maehr, R. F. Noss, and J. L. Larkin, 205–224. Island Press, Washington, DC.

Schaller, D. 1996. "The Ecocenter and tourist attraction: Ely and the International Wolf Center." Department of Geography, University of Minnesota, Minneapolis. www.eduweb.com/schaller/IWCSummary.html.

Shinneman, D., R. McClellan, and R. Smith. 2000. *The State of the Southern Rockies Ecoregion.* Southern Rockies Ecosystem Project, Nederland, CO.

Smith, D. W., R. O. Peterson, and D. B. Houston. 2003. "Yellowstone after wolves," *BioScience* 53:330–340.

Southern Rockies Ecosystem Project. 2000. *Summary of Base Data and Landscape Variables for Wolf Habitat Suitability on the Vermejo Park Ranch and Surrounding Areas.* Final Report to the Turner Endangered Species Fund, Bozeman, MT.

Stenlund, M. H. 1955. *A Field Study of the Timber Wolf (Canis lupus) on the Superior National Forest, Minnesota.* Technical Bulletin No. 4, Minnesota Department of Conservation. St. Paul, MN

Terborgh, J., J. A. Estes, P. Paquet, K. Ralls, D. Boyd-Heger, B. J. Miller, and R. F. Noss. 1999. "The role of top carnivores in regulating terrestrial ecosystems," in *Continental Conservation: Scientific Foundations for Regional Reserve Networks,* ed. M. E. Soulé and J. Terborgh, 39–64. Island Press, Washington, DC.

Terborgh, J., L. Lopez, P. Nuñez V., M. Rao, G. Shahabuddin, G. Orihuela, M. Riveros, R. Ascanio, G. H. Alder, T. D. Lambert, and L. Balbas. 2001. "Ecological meltdown in predator-free forest fragments," *Science* 294:1923–1925.

Thiel, R. P. 1993. *The Timber Wolf in Wisconsin: The Death and Life of a Magnificent Predator.* University of Wisconsin Press, Madison.

Theobald, D. M., H. Gosnell, and W. E. Riebsame. 1996. "Land use and landscape change in the Colorado mountains. II: A case study of the East River Valley, Colorado," *Mountain Research and Development* 16:407–418.

United States Ninth Circuit Court of Appeals. 2001. United States Court of Appeals for the Ninth Circuit, Orders Nos. 99-56362, 00-55496; D.C. No. CV-97-02330-TJW/LSP.

U.S. Fish and Wildlife Service (USFWS). 1994. *The Reintroduction of Gray Wolves to Yellowstone National Park and Central Idaho.* Final Environmental Impact Statement, Denver, CO.

———. 1996. *Reintroduction of the Mexican Wolf Within Its Historic Range in the Southwestern United States.* Final Environmental Impact Statement, Albuquerque, NM.

———. 1998. "Establishment of a nonessential experimental population of the Mexican grey wolf in Arizona and New Mexico," *Federal Register* 63:1752–1772.

———. 2000. *Grizzly Bear Recovery in the Bitterroot Ecosystem.* Final Environmental Impact Statement, Denver, CO.

———. 2003a. "Final rule to reclassify and remove the gray wolf from the list of endangered and threatened wildlife in portions of the conterminous United States," *Federal Register* 68:15804–15875.

———. 2003b. "Removing the eastern distinct population segment of the gray wolf from the list of endangered and threatened wildlife," *Federal Register* 68:15876–15879.

———. 2003c. "Removing the western distinct population segment of the gray wolf from the list of endangered and threatened wildlife," *Federal Register* 68:15879–15882.

Varley, J. D., and W. G. Brewster, eds. 1992. *Wolves for Yellowstone? A Report to the United States Congress,* vols. 3 and 4. National Park Service, Yellowstone National Park.

Weaver, J. L. 1978. *The Wolves of Yellowstone.* Natural Resources Report No. 14. National Park Service, Government Printing Office, Washington, DC.

Weber, W., and A. Rabinowitz. 1996. "A global perspective on large carnivore conservation," *Conservation Biology* 1046–1054.

Webster, D. 1999. "Welcome to Turner country," *Audubon* 101:48–56.

Yellowstone National Park, U.S. Fish and Wildlife Service, University of Wyoming, University of Idaho, Interagency Grizzly Bear Study Team, and University of Minnesota Cooperative Park

Studies Unit. 1990. *Wolves for Yellowstone? A Report to the United States Congress*, vols. 1 and 2. National Park Service, Yellowstone National Park.

Yellowstone Park Foundation. 2003. "Wolf economic study. Yellowstone Park Foundation." Unpublished report, Bozeman, Montana.

Young, S. P. 1970. *The Last of the Loners.* Macmillan, New York.

Young, S. P, and E. A. Goldman. 1944. *The Wolves of North America.* American Wildlife Institute, Washington, DC.

Conclusion

Nina Fascione, Aimee Delach, and Martin E. Smith

Maintaining viable populations of predators in landscapes that are increasingly dominated by humans is one of the foremost challenges of wildlife conservation in the twenty-first century. There is growing knowledge of the ecological benefits provided by carnivores, as well as increasing recognition that strong public interest in predators helps drive tourism dollars to places where predators reside. However, real conflicts stand in the way of carnivore recovery and must be overcome if predators and people are to coexist into the future. Human activities affect carnivores in ways that can be detrimental to a single individual or to the viability of entire populations and species. Predators impact people as well. The costs associated with carnivore conflict—including both harm caused by carnivores, and also the costs of measures to mitigate harm to carnivores—are real and can be substantial, in some cases posing significant financial hardship to the affected parties. Both types of conflicts occur across various landscapes, including rural, urban, exurban, and political.

Fortunately, it is possible to prevent or ameliorate many of the conflicts between predators and people in shared landscapes. In rural areas, where most of the problems involve livestock, conflicts can be minimized by altering animal husbandry practices and by aversive conditioning of predators. Failing that, compensation programs can reimburse landowners for losses and shift the financial burden of predator recovery from local landowners to carnivore advocates (private programs) or the broader public. Solutions also exist to reduce conflicts in more developed areas. Wildlife road crossings can reduce vehicle strikes, updated power line design can minimize electrocutions of raptors, and fairly simple solutions exist to reduce problems such as garbage intrusions.

In many cases, the potential solutions are known but require the funding and motivation for implementation. However, in some cases, predator-human conflicts are interwoven with issues of land use, water allocation, and invasive species; solutions, while achievable, will require considerably bigger changes and concordantly larger amounts of political will. For example, the San Joaquin kit fox lives in habitats that have been fragmented by an agriculture industry that is central to the human economy across its range. Furthermore, humans have introduced to the landscape red foxes, which compete with the kit fox, and artificial water bodies that further enhance the red fox's advantage. Preserving the kit fox in the face of these numerous challenges will require diligence and creativity. "Safe harbor" agreements that provide incentives for landowners to protect suitable kit fox habitats are one way that conservationists have fostered acceptance of this small, threatened carnivore.

The contributors to this volume have demonstrated that while much research has been done on conflicts between carnivores and humans, many questions remain unanswered. More data are needed on the causes and patterns of livestock depredations, and on the efficacy of various livestock husbandry techniques and deterrent devices. For instance, what ecological, spatial, and behavioral cues trigger predators to prey on livestock? Can we predict with some degree of certainty when and where livestock is most at risk for depredation, and so focus deterrent measures most efficiently? How do multiple predator species that share the landscape partition prey resources, both natural and livestock, and how does the presence of multiple carnivore species impact both depredation patterns and ranchers' perceptions of the causes of depredation? What is the most effective deterrent or combination of husbandry practices and avoidance mechanisms, and how should these be implemented? Do the answers to these questions vary spatially, seasonally, with different predator species, and in response to changing dynamics between carnivores and their natural prey species?

Likewise, more research is needed to understand conflict patterns and solutions in more human-dominated landscapes. As we have seen, basic ecological research on diets, habitat use, and niche partitioning among urban carnivores can help managers more accurately predict when and where conflicts will occur, and what must be done to prevent or minimize these. It is also important to learn more about how various facets of human society—roads, buildings, power lines, garbage dumps, chemicals, introduced species, etc.—impact carnivores, and what practices and technologies will make these aspects of human presence more compatible with carnivore conservation. Furthermore, as new technologies emerge, carnivore research and management will need to keep pace so that managers can identify and handle conflicts that might arise. Wind turbines, while a promising source of clean energy, are an example of a technology with the potential to impact carnivores directly, via raptor mortality, and indirectly, through habitat modification. Conservationists must work hand in hand with engineers, planners, and others to avoid and minimize impacts from these technologies. Research will also help us understand which habitats are most important to conserve as human activities encompass ever more of the landscape, and how best to preserve and connect the fragments of habitat that remain.

There is also a tremendous need for more information about the social and economic impacts of predators. Few hard data exist on the impact of predator-related tourism to local economies, limiting us in our predictive power to weigh the costs and benefits, for instance, of predator recovery in communities that derive significant income from ungulate hunting. Will returning predators to these areas harm these economies by reducing, in reality or in perception, hunter success, or will the presence of carnivores generate tourism interest that compensates for any decrease in hunting-related dollars?

Obviously, research of the type and scope described here represents a significant financial commitment. An even greater commitment will be needed to fully implement the recommendations presented in this volume for reducing conflict between people and predators. Unfortunately, money for predator research tends to reside at the uncomfortable end of the funding hierarchy: funding for predators often competes with other nongame projects, which frequently take a back seat to game-related management, and tight state and federal budgets tend to put a squeeze on research in general and natural resource funding in particular. Predator advocates need to continue to "think outside the box" for funding from private as well as public sources. Financial backing will need to come from the full gamut of options, including government agencies, nonprofit dollars supported by the general public, and corporate sponsorship.

Conservationists may be squeamish about corporate partnerships—rightfully worried about assisting in green-washing schemes. But there are numerous cases in which corporate support can provide much needed funding for research, education, and conservation. For example, automaker Jaguar North America formed the Jaguar Conservation Trust in 2003 to help preserve the company's namesake. And, perhaps more controversially, Exxon has financially supported large cat conservation through their Save the Tiger Fund. Any corporation or sports team that uses predators as its mascot or logo should feel an obligation to help preserve the species it uses to create an image and sell products. Thus the Chicago Bears, Detroit Lions, Carolina Panthers, and Minnesota Timber Wolves could all be potential funding sources.

On a larger scale, carnivore expert David Macdonald has proposed what he calls "the Biodiversity Impacts Compensatory Scheme" (Gittleman 2001), in which corporations causing unavoidable environmental damage would be obligated to pay into an environmental mitigation fund. This money would then be used to fund global biodiversity conservation efforts. Defenders of Wildlife is working with a coalition of ranchers to determine the viability of an insurance program, tied to a community foundation, that would provide compensation for livestock depredations as well as funds for other community needs.

But perhaps most important, we need the political will to find and implement solutions to reduce conflicts and preserve and restore carnivore populations. Past conflicts and the potential for future conflicts have engendered antipredator sentiments in many areas of prime carnivore habitat. These can be a strong motivating force in local politics, as evidenced by the numerous antipredator legislation and ordinances proposed in counties and states located in potential predator habitat. Furthermore, since most predator recovery to date has taken place under the auspices of the federal Endangered Species Act, wolves and other predators to some connote regulation, government interference, and loss of property rights.

Without significant investment of time and resources into changing the political climate and building the will to coexist with predators, it will be very difficult to advance the argument that funding the research and implementation of solutions is a wise investment. Paradoxically, however, it is likely impossible to build such political will without demonstrating that conflicts can be prevented, mitigated, or resolved. Building political will also means convincing those who could face hardship from predator conflicts that there are advantages to living with, and not eradicating, carnivores. This likely means the development of incentives, such as green certification of products and businesses that undertake the extra effort to coexist with predators. Predator advocates need to take the lead in finding innovative ways to manage conflicts, in developing incentives to increase the attractiveness of coexisting with predators, and in seeking new sources of funding to make these things possible.

Individual citizens can often lead the way in developing solutions to conflicts. Although large-scale carnivore restoration will require coordinated efforts between government agencies, nongovernmental agencies, and the general public, individuals can lead by example. In the southwest United States, ranchers Will and Jan Holder turned to an unconventional method of keeping their family ranch solvent during difficult economic times. The Holder ranch, in southern Arizona, is home to several predator species, including mountain lions, coyotes, black bears, and newly restored Mexican wolves. Ironically, Will Holder's great grandfather is believed to have killed the last Mexican wolf in the state. Yet the Holders use a variety of the nonlethal methods described in Part 1 of this volume to control predators on their property, and they market their products with a "predator-friendly" label through Ervin's Grassfed Beef. This "ecolabel" allows the Holders to market their predator-friendly beef for premium prices at food cooperatives and health food stores in various parts of the country, as well as one of New York City's finest restaurants.

The "ecolabel" program is not without its challenges—for example, marketing is difficult and costly for the individual rancher. Another challenge is the kind of reaction the Holders received from their neighbors, who were initially less enamored of maintaining healthy predator populations. Fortunately, the Holders are creative and persistent, and their ability to command a premium price for a standout product has helped them stay in business at a time when many other ranches have gone under. Interestingly, some of the Holders' neighbors have recently become interested in the predator-friendly ranching idea.

This example can teach us something about the future of carnivores in a populated world. For people and predators to coexist in the future we must look not only at how to manage predators but at how to manage our own attitudes and behaviors. The Holders chose to modify their ranching practices rather than fight a difficult battle to control the predators they share the land with. The results have benefited the Holders, the region's predators, and the larger ecosystem. All of us can learn from this.

Whether we are urbanites or rural dwellers, we have an obligation to contribute to the biological diversity that will ensure all of us and our progeny a healthy environment and future. Predators are a key component of our natural world. We owe it to ourselves and our ecosystems to devote our full creative and fiscal resources to resolving conflicts. With a little willingness and ingenuity, people and predators can achieve peaceful coexistence.

Literature Cited

Gittleman, J. L., S. M. Funk, D. Macdonald and R. K. Wayne, eds. 2001. *Carnivore Conservation*. Cambridge University Press, Cambridge.

About the Editors

Aimee Delach is the senior program associate for species conservation at Defenders of Wildlife.

Nina Fascione is the vice president for species conservation at Defenders of Wildlife.

Martin E. Smith is the former carnivore biologist at Defenders of Wildlife.

About the Contributors

Tina Arapkiles is the Southwest regional representative for the Sierra Club.

Clint W. Boal is a researcher with the USGS-BRD Texas Cooperative Fish and Wildlife Research Unit at Texas Tech University in Lubbock.

Stewart W. Breck is a research scientist with the USDA Wildlife Services National Wildlife Research Center and has several research projects focused on predator-prey interactions, carnivore ecology and management, and human-wildlife conflict. He has worked on the conservation and management of a variety of species, including wolves, black-footed ferrets, and black rhinos, and the ecology of beaver and small desert rodents.

Brian Brost is a University of Wisconsin student studying wolf pack viability in relation to habitat quality.

Carolyn Callaghan is the executive director and director of research with the Central Rockies Wolf Project.

Howard O. Clark Jr. earned a Master's degree in biology at California State University, Fresno in 2001. His thesis investigated the interference and exploitation competition between San Joaquin kit foxes and red foxes. He worked as a wildlife biologist for seven years at California State University, Stanislaus' Endangered Species Recovery Program. He is currently employed as a mammalogist at H. T. Harvey & Associates, in Fresno, California.

Brian L. Cypher is a research ecologist and coordinates several of the Endangered Species Recovery Program's research projects on San Joaquin kit foxes.

About the Contributors

Rob Edward is the director of carnivore restoration at Sinapu in Boulder, Colorado.

James A. Estes is a research biologist with the USGS Western Ecological Research Center in Santa Cruz, CA. He is an international expert on sea otters and their influence on marine ecosystem function. He has published nearly 70 scientific articles and reports on wildlife ecology, predation, and conservation.

C. Cormack Gates is an associate professor of environmental science in the Faculty of Environmental Design at the University of Calgary.

Stanley D. Gehrt is an assistant professor with the School of Natural Resources at Ohio State University. His research interests include urban ecology of mammals, behavioral ecology of carnivores, and infectious diseases in wildlife.

David E. Grubbs is an associate professor of biology at California State University, Fresno.

Patrick A. Kelly is the Coordinator and Director of the Endangered Species Recovery Program, is an associate professor of biological sciences at California State University, Stanislaus, and is an adjunct associate professor of biology at California State University, Fresno.

David S. Maehr is an associate professor of conservation biology in the Forestry Department of the University of Kentucky. He is the author of *The Florida Panther: Life and Death of a Vanishing Carnivore* (Island Press 1997) and over thirty articles and papers on the Florida panther. His research interests also include elk restoration in Kentucky, conservation of Florida black bears, and the role of university lands in conservation of biological diversity.

R. William Mannan is a professor with the University of Arizona's School of Renewable Natural Resources, Tucson. His research specializes in the ecology of birds of prey in urban landscapes, particularly Harris's hawks, Cooper's hawks, burrowing owls, and elf owls in Tucson.

David J. Mattson is a research wildlife biologist with the USGS Forest and Rangeland Ecosystem Science Center, Colorado Plateau Field Station. He has studied the ecology and conservation of grizzly bears for twenty-one years, as well as the sociopolitical aspects of carnivore conservation.

Tyler Muhly is a student in the Faculty of Environmental Design at the University of Calgary. His research is focused on site depredation analysis.

Marco Musiani is an adjunct professor at Prescott College, Prescott, Arizona, and is currently based at the University of Rome, Italy. There, the Na-

About the Contributors

tional Sciences and Engineering Research Council of Canada is supporting his project on the relationships between wolves, dogs, and humans in the context of high levels of depredation on livestock.

Martin Nie is an assistant professor of natural resource policy at the University of Montana College of Forestry and Conservation in Missoula.

Christopher M. Papouchis is the conservation biologist for the Mountain Lion Foundation in Sacramento, California.

Michael K. Phillips is the executive director of the Turner Endangered Species Fund. He has conducted wildlife research, with an emphasis on large carnivores, throughout the United States, Alaska, and Australia. His professional interests include conservation and restoration of imperiled species, integration of private land in conservation projects, and privatization of endangered species recovery programs.

Bill Ruediger is the ecology program leader for highways with the USDA Forest Service in Missoula, Montana. He has researched the effects of habitat connectivity and landscape linkages with respect to conservation of grizzly bears, wolverine, lynx, fisher, and trout and salmon rivers.

Suzanne Stone is the Rocky Mountain field representative at Defenders of Wildlife.

Elisabetta Tosoni is a graduate student at the University of Rome, La Sapienza.

Adrian Treves is with the Center for Applied Biodiversity Science at Conservation International in Madison, Wisconsin. His research interests include carnivores, human-carnivore conflict, primate behavior, and ecotourism.

Gregory D. Warrick is a wildlife biologist and manager at the Center for Natural Lands Management, a non-profit organization for the protection and management of natural resources.

Jane E. Wiedenhoeft is with the Wisconsin Department of Natural Resources.

Daniel F. Williams is the executive director of the Endangered Species Recovery Program, and is a retired Associate Professor of Biological Sciences at California State University, Stanislaus.

Adrian P. Wydeven is the head of the wolf recovery program at the Wisconsin Department of Natural Resources.

Index

Accipiter cooperii. See Cooper's hawk
Acinonyx jubatus (cheetahs), 151
Adaptive management: description, 15; "learning by doing," 22; livestock depredation research and, 22–23
African wild dogs *(Lycaon pictus)*, 157
Ailuropoda melanoleuca (giant panda), 151–152
Alaska Board of Game (BOG): criticism of, 202; membership in, 200; responsibilities of, 200–201; wolf management by, 205, 208, 209
Alaska Department of Fish and Game (ADFG): funding for, 201; responsibilities of, 200–201
"Alaskans Against Snaring Wolves," 205
Alaska Wildlife Alliance, 202
Alaska wolf management: airborne wolf hunting, 205; lessons learned summary, 206–209; predator-prey dynamics and, 207–208; stakeholder-based approach, 208–209; state responsibility and, 205; wolf-related ballot initiatives, 205–206
Alberta Conservation Association, 53
Alces alces (moose), 54
Alligator River National Wildlife Refuge, 254
Alopex lagopus semenovi (Mednyi Island arctic fox), 157
Alvarez, K., 190
Andean mountain cat *(Oreailurus jacobitus)*, 159

Animal husbandry: compensation programs and, 66; livestock depredation and, 64, 65–66, 233–234, 263; livestock depredation research, 18–20
Anne, K.E., 226
"Anthropogenic ecological traps," 110
Antilocapra americana (pronghorn antelope), 54
Aquila chrysaetos (golden eagles), 4
Arapkiles, T., 178
Aspen and elk/wolves, 3–4

Badger *(Meles meles)*, 155
Badger *(Taxidea taxus)*, 5, 144
Baker, R.O., 99
Bald eagles *(Haliaetus leucocephalus)*: recovery of, 6; Yellowstone wolves and, 4
Ballot initiatives: challenges/limitations of, 204–205; historical overview, 203–204; on hunting/trapping practices, 204; mountain lions and, 232; state wildlife commissions and, 203; urban vs. rural residents and, 205; wolf-related initiatives, 205–206, 207, 208, 210
Banff National Park: highways/carnivores and, 135, 138, 144; wildlife crossings of, 144
Baron, D., 226
Bat-eared fox *(Otocyon megaiotis)*, 157
Baylisascaris procyonis, 84
Beier, P., 225–226
Belding's savannah sparrow *(Passerculus sandwichensis beldingi)*, 127

Bennett, L.E., 249–250
Big Cypress National Preserve, 180, 181, 189
Bighorn sheep *(Ovis canadensis)*, 54
Birds of prey: prey at forest edges, 156–157. *See also* Urban landscapes/birds of prey
Bison *(Bison bison)*, 54
Black bear *(Ursus americanus)*: highways and, 132, 133, 136, 139, 141, 143, 144; wildlife crossings and, 144
Black-footed ferret *(Mustela nigripes)*, 156, 157, 160
Blue Range Wolf Recovery Area, 243, 244
Boal, C.W., 78
Bobcat *(Lynx rufus)*, 157
Branta canadensis (Canada geese), 85
Breck, S., 10
Brown bear. *See* Grizzly bear *(Ursus arctos)*
Buteo jamaicensis (red-tailed hawk), 106
Buteo swainsoni (Swainson's hawk), 108

Cactus ferruginous pygmy-owl *(Glaucidium brasilianum cactorum)*, 112
Caddick, G.B., 179
California least tern *(Sterna antillarum browni)*, 127
California light-footed clapper rail *(Rallus longirostris levipes)*, 127
Canada geese *(Branta canadensis)*, 85
Canada lynx. *See* Lynx *(Lynx canadensis)*
Canis aureus (jackal), 151, 154, 163
Canis latrans. See Coyote
Canis lupus. See Wolf
Canis lupus baileyi (Mexican wolf), 197, 243, 252
Canis lupus lupus (Italian wolf), 158
Canis lupus signatus, 163
Canis simensis (Ethiopian wolves), 157, 159
Carnivore endangerment: conceptual models for, 168–172; of generalists, 163, 164, 165; of large species, 164, 165, 168, 169, 170, 173; of specialists, 163, 164, 165, 168, 171–172, 173; summary tables, 166–167; taxonomy, 164–168
Carnivore endangerment/biological factors: behavioral flexibility, 163, 164, 165, 168; body size, 162–163, 164, 165, 168; range size, 163
Carnivore endangerment/human factors: collision with vehicles, 155–156; conceptual models of, 168–172; deforestation, 156, 159, 160; disease, 157; distal factors, 157–162; harvest of body parts, 153–154, 160; human/carnivore joint concentrations, 160–161; human density, 158–159; human values/perspectives, 161–162, 172–173; loss of cover/hunting habitat, 156–157; loss of prey, 156; overview, 172–174; pollution, 160; poverty and, 159, 172; proximal factors, 153–157; retaliation for depredation, 154–155; road density and, 157–158, 161, 172; wealth and, 159–160, 172
Carnivores: attacks on humans, 1, 162, 225–226; cultural heritage and, 2–3; gender/conflicts, 31; historical heritage and, 2–3; injured individuals/conflicts, 29, 31; island biogeography theory and, 152–153
Carnivores (large): endangerment of, 164, 165, 168, 169, 170, 173; habitat requirements of, xi–xii, 5; human persecution of, xi, 5, 9; impacts on humans (overview), 1; role in ecosystem (overview), xii, 1, 2–4; as vulnerable, 5. *See also specific carnivores*
Carroll, C., 251, 252
Catopuma temminckii, 156
Cause-and effect relationship, 14
Cerocyon thous (crab-eating fox), 157
Chase, L.C., 232
Cheetahs *(Acinonyx jubatus)*, 151
CITES (Convention on International Trade in Endangered Species), 168
Clark, H.O., Jr., 78
Clark, T., 199
Clevenger, T., 135
Clouded leopard *(Neofelis nebulosa)*, 154, 156
Coexistence (predators/people): corporate partnerships, 265–266; importance of, xii, xiii; new technologies and, 265; overview, 263–268; research funding, 265–266; research needed, 264–266; as two-way street, 6
Colorado Wildlife Commission, 242
Columbina inca (Inca dove), 110
Compensation programs: animal husbandry efforts and, 66; community foundations and, 266; costs of, 58, 60; Defenders of Wildlife and, 53–54, 266; economic burden and, 66; in western North America, 53–54; in Wisconsin, 45

Compound 1080, 127
Conservation: amount of Earth protected, 5; island biogeography theory and, 152–153; knowledge/skills and, 173–174; paper parks and, 5; protection limits, 152–153
Conservation biology goals, xi
Conservation Breeding Specialist Group (CBSG), 250
Conservation easements, 145, 146
Control (research), 14
Convention on International Trade in Endangered Species (CITES), 168
Cooper's hawk *(Accipiter cooperii)*: collision with windows/vehicles, 110; relocation of nests, 114; trichomoniasis in, 110, 113; urban home ranges, 108; urban landscape resources, 108; urban nesting sites, 106; urban poisoning of, 110; urban population dynamics of, 105, 110–111
Copeland, J., 139
Corvus corax (raven), 4
Coyote *(Canis latrans)*: attacks on humans, 84, 99; attacks on pets, 84, 85, 91–92; behavioral flexibility of, 163; breeding pairs/livestock depredation, 29, 44; cats as prey, 84, 85, 91–92; coevolution with kit foxes, 125–126; cultural inheritance and, 98; diet of, 91–92; exclusion of red foxes by, 126, 127–128; highways and, 132, 136, 141, 144; home range size of, 93, 94–95, 96; hunting coyotes with dogs, 35, 43; increase in numbers, 151; livestock depredation and, 155; nuisance issues with, 82–83; poison control banning, 66; poisoning of, 120; predation on white-tailed deer, 85; problem individuals with, 16; public health issues with, 83–84; red foxes and, 122, 126, 127–128; wildlife crossings and, 144; wolves and, 4, 125. *See also* Urban landscapes/skunks, raccoons, coyotes
Crab-eating fox *(Cerocyon thous)*, 157
Craighead, L., 142
Cynomys spp. (prairie dog), 156, 163

DDT ban, 6
Decker, D., 223
Defenders of Wildlife, 53–54, 266
Deforestation, 156, 159, 160
Denali National Park, 43
Distemper, 157

Distinct population segments (DPS), 241, 242–243
Dogs: depredation of, 31, 35–43, 41, 44, 46, 64; guardian dogs, 66–67, 233, 234; hunting bear with, 43, 209; hunting coyotes with, 35, 43; hunting mountain lions with, 224, 228; transmission of disease by, 157
DPS (distinct population segments), 241, 242–243

Earle, B.D., 126
Eastern screech-owl *(Megascops asio)*, 105, 107
"Ecolabel" program, 267
Edge habitat, 122
Edward, R., 178
Electric shock collars, 45, 67
Elk *(Cervus elaphus)*: foraging/predators and, 3–4; highways and, 139; as wolf prey, 54
Endangered Species Act (ESA): development of land and, 112, 114; following reintroduction/recovery, 197–198; grizzly bear listing, 5; species recovery with, 6, 266
Enhydra lutris (sea otter), xii, 2, 3
Environmental Defense, 128
Environmental impact statements (EIS), 213, 214
Erethizon dorsatum (porcupine), 156
ESA. *See* Endangered Species Act
Ethiopian wolves *(Canis simensis)*, 157, 159
Eurasian lynx *(Lynx lynx)*, 154
European brown bear *(Ursus arctos arctos)*, 163
Everglades National Park, 180, 181
Extinction rates, 5
Exxon, 266

Falco columbarius (merlin), 105
Falco peregrinus. *See* Peregrine falcon
Felis catus (housecat), 84, 85, 91–92
Felis pardina (Iberian lynx), 135
Fencing techniques, 67
Ferreras, P., 135
Fisher *(Martes pennanti)*: forest cover and, 156; fur harvesting, 154; highways and, 133, 141, 144; population counts of, 5; wildlife crossings and, 144
Fladry effectiveness/research, 20–21, 24, 69–70
Flagship species, 4

278 Index

Florida black bear *(Ursus americanus floridanus)*, 155, 187
Florida Fish and Wildlife Conservation Commission (FFWCC), 189
Florida panther *(Puma concolor coryi)*: differences with other cougar populations, 180; habitat loss, 180, 223; highway mortality and, 155, 184–185, 186; highways and, 132, 135–136, 143, 155, 184–185, 186; numbers of, 184; panther recovery vs. development, 191; politics and, 187–193; private landowners and, 189; recovery/dispersal summary, 193–194; recovery plan for, 189–190; south-central Florida and, 184–185, 192; spatial constraint of, 180–181; Texas cougars and, 183, 184, 186, 223; translocation and, 188, 189
Florida panther *(Puma concolor coryi)* dispersal: critical habitat and, 192; description, 180–182, 182–187; disruption of, 180–181; gender differences with, 186–187; importance of, 179–180; mortality with, 183, 184, 185; resistance of landscape and, 182, 183, 185; source population and, 182, 183, 185
Fur harvesting, 154

Gallagher, P.B., 136
Gehrt, S.D., 78
Generalists endangerment, 163, 164, 165
Giant panda *(Ailuropoda melanoleuca)*, 151–152
Gibeau, M.L., 136
Gill, B., 202–203
Glaucidium brasilianum cactorum (cactus ferruginous pygmy-owl), 112
Golden eagles *(Aquila chrysaetos)*, 4
Gould, G.I., 122
Green certificates, 266–267
Grinnell, J., 221
Grizzly bear *(Ursus arctos)*: attacks on humans, 1; ESA listing of, 5; extirpation/persecution of, 151, 160–161, 172, 220; highways and, 132, 133, 136, 139, 140, 141, 144; livestock depredation and, 1, 28, 154; population of, 5; problem individuals with, 16; translocation program for, 140; as umbrella species, 4; wildlife crossings and, 144; Yellowstone wolves and, 4

Guardian dogs: mountain lion livestock depredation and, 233, 234; wolf livestock depredation and, 66–67
Gulo gulo. See Wolverine

Habitat: edge habitat, 122; Florida panther and, 180, 192, 223; fragmentation/loss with highways, 137–141, 146–147; habitat suitability study/SRM, 250, 251–252; human expansion and, 2, 4, 118, 133; large carnivore requirements, xi–xii, 5; loss of cover/hunting habitat, 156–157; mountain lion fragmentation/loss of, 224–225, 231; PHVA, 250–251; San Joaquin kit fox and, 119–120, 124–125, 128; in urban landscape, 6; wildlife habitat linkage analysis, 142–144, 147
Habitat connectivity: importance of, 146–147; overview, 142–144
"Habitat imprinting," 107
Hadidian, J., 97–98
Haliaetus leucocephalus. See Bald eagles
Harris, L.D., 136
Harris hawk *(Parabuteo unicinctus)*, 111
Harvest of body parts, 153–154, 160
Helarctos malayensis (sun bear), 151
Heuer, K., 136
Hibben, F., 221
Highways/carnivores: acquisition of wildlife linkage areas, 145–146; agency policies, 146–147; associated human development, 141–142; background, 132–133; carnivore sensitivity to highways, 133–135; conclusion, 147–148; differential avoidance/recovery, 141; direct mortality, 135–136, 155–156, 157–158, 161; dispersal barriers, 140; fencing highway right-of-ways, 145; habitat fragmentation, 137–141, 146–147; habitat loss, 136–137, 146–147; home range constriction, 139; road densities and, 157–158, 161, 172; roadless areas/development, 138; seasonal movement disruption, 139–140; solutions, 142–148, 156; transportation agencies and, 146, 147; U.S. highway system map, 138; vehicle accidents and, 143, 144; wildlife crossings, 144, 156, 263; wildlife habitat linkage analysis, 142–144, 147
Hjermann, D.O., 182

Index   279

Hoctor, T.S., 189
Holder, Will/Jan, 267
Hornocker, M., 221–222
Housecat *(Felis catus)*, 84, 85, 91–92
Howard, W.E., 182
Humane Society of the United States (HSUS), 202
Humans: California population, 118, 125; as carnivore, 4; common patterns in carnivore conflicts, 28–29, 30; Florida population, 180; habitat expansion of, 2, 4, 118, 133; human density/carnivore endangerment, 158–159; natural resource consumption by, 4–5; present/future population, xi, 4, 133, 134–135; Rocky Mountain area population, 134–135

Iberian lynx *(Felis pardina)*, 135
Ictinia mississippiensis. *See* Mississippi kite
Idaho's Wolf Conservation and Management Plan, 210–211, 214
Idaho Wildlife Federation, 203
Ims, R.A., 182
Inca dove *(Columbina inca)*, 110
Interagency Grizzly Bear Committee, 142
International Wolf Center, 254
Italian wolf *(Canis lupus lupus)*, 158

Jackal *(Canis aureus)*, 151, 154, 163
Jackson, P., 225
Jaguar *(Panthera onca)*, 5, 29
Jaguar Conservation Trust, 266
Jaguar North America, 266
Jamison, E.V., 52
Johnson, D.H., 24

Keystone species, 2, 121
Krebs, J., 136

"Learning by doing," 22
Leishmaniasis, 157
Leopard *(Panthera pardus)*, 29, 151
Leopardus tigrinus (oncilla), 156
Leopold, A., 221
Leptospirosis, 83
Leptospirosis interrogans/autumnalis, 83
Lepus americanus, 156
Lewis, D., 136
Linnell, J.D.C., 16, 31

Lion *(Panthera leo)*, 29, 151
Livestock depredation: carcass removal and, 18–19; common patterns with, 28, 30, 31; economic dimension of, 13, 51; emotional dimension of, 13, 51; by grizzly bears, 1; "hot spots" of, 15–16; by mountain lions, 1; overview, 9–10, 28; solutions for, 18–25, 44–46, 66–68, 69–70, 155, 173, 233–234. *See also* Animal husbandry; Wolf *(Canis lupus)*/depredations in western North America; Wolf *(Canis lupus)*/depredations in Wisconsin
Livestock depredation research: adaptive management and, 22–23; animal husbandry and, 18–20; commitment to learning, 23; difficulties with, 13–14, 21–22; "hot spots," 15–16; importance of, 13; international oversight committee for, 25; overview, 24–25; problem individuals and, 16–17; replication importance, 23–24; tools for managing carnivores, 20–21
Living with Lions program, 234
Logan, K., 226, 227, 234, 235
Lutra canadensis (river otter), 141, 154
Lutra lutra, 154
Lycaon pictus (African wild dogs), 157
Lynx *(Lynx canadensis)*: fur harvesting, 154; highways and, 132, 133, 135, 139–140, 141; livestock depredation and, 154; loss of prey, 156; population decline, 5; problem individuals with, 16; translocation program for, 140
Lynx lynx (Eurasian lynx), 154
Lynx rufus (bobcats), 157

Macdonald, D., 266
Maehr, D.S., 136, 177–178, 179
Magpie *(Pica hudsonia)*, 4
Malthus, xi
Management paradigm, 201
Manfredo, M.J., 253
Mange, 157
Manipulative experiments: description, 14; difficulties with, 14
Mannan, W.R., 78
Marten *(Martes americana)*: forest cover and, 156; fur harvesting, 154; highways and, 132, 141, 144; wildlife crossings and, 144
Marten *(Martes martes)*, 154

Martes pennanti. See Fisher
Mattson, D.J., 79
McCoy, E.D., 191
Mech, L.D., 16, 244–245
Mednyi Island arctic fox *(Alopex lagopus semenovi)*, 157
Megascops asio (eastern screech-owl), 105, 107
Meles meles (badger), 155
Mephitis mephitis. See Skunks, striped
Merlin *(Falco columbarius)*, 105
Mesocarnivores, 85. *See also individual species*
Metareplication, 15
Mexican wolf *(Canis lupus baileyi)*, 197, 243, 252
Mink *(Mustela lutreola)*, 154
Mink *(Mustela spp.)*, 144
Mississippi kite *(Ictinia mississippiensis)*: diving at people by, 111; shooting of, 111; urban landscape resources, 108; urban nesting sites of, 106–107; urban vs. rural areas, 105
Mitigation, 137, 146
Montana: actions following wolf recovery, 213–214, 241–242; Environmental Policy Act (MEA), 213; Wolf Conservation and Management Plan, 213–214; Wolf Management Advisory Council, 213
Moose *(Alces alces)*, 54
Mountain Lion Foundation, 234
Mountain lions *(Puma concolor)*: attacks on humans, 1, 225–226; ballot initiatives and, 232; cats killed summary, 227; changing attitudes towards, 220, 221–222, 223–224; current management trends, 226–229; ecological role of, 219, 229, 230; habitat loss/fragmentation and, 224–225, 231; highways and, 139, 141, 144, 225; human threats to, 224–225, 226–229; knowledge of, 219–220, 234–235; livestock depredation and, 1, 28, 154, 155, 229, 233–234; management/human attitudes and, 219–220; nonlethal conflict resolution, 233–234; persecution of, 220–222; problem individuals with, 16; sport hunting of, 224, 226–229; status in U.S./Canada, 222–223; threats to populations, 224–226; translocation of, 233; wildlife crossings and, 144. *See also* Florida panther

Mountain lions *(Puma concolor)* conservation: collaborative networks development, 232–233; current conservation, 229–230; landscape-level strategies, 230–232; overview, 235; research and, 234–235
Mourning dove *(Zenaida macroura)*, 110
Mule deer *(Odocoileus hemionus)*, 54
Murphy, D.D., 137
Musiani, M., 10, 21, 24
Mustela lutreola (mink), 154
Mustela nigripes (black-footed ferrets), 156, 157, 160
Mustela putorius (polecat), 154
Mustela spp. (mink), 144
Mustela spp. (weasel), 144, 151
Muth, R.M., 52

National Geographic Society, 221
Ned Brown Forest Preserve, 85–87. *See also* Urban landscapes/skunks, raccoons, coyotes
Neofelis nebulosa (clouded leopard), 154, 156
New Mexico Wildlife Commission, 242
Nie, M., 178
Nonconsumptive interest groups: funding and, 214–215, 216; state wildlife commissions and, 200, 201–203, 207–208, 209
"Northern Rocky Mountain Grizzly Bear and Gray Wolf Management Trust," 215
Nowell, K., 225

Observational studies, 14
Odocoileus hemionus (mule deer), 54
Odocoileus virginianus (white-tailed deer), 54, 85
Okeechobee, Lake, 180
Oncilla *(Leopardus tigrinus)*, 156
Oreailurus jacobitus (Andean mountain cat), 159
Otocyon megaiotis (bat-eared fox), 157
Ovis canadensis (Bighorn sheep), 54

Pacelle, W., 204
Panthera leo (lion), 29, 151
Panthera onca (jaguar), 5, 29
Panthera pardus (leopard), 29, 151
Panthera tigris (tiger), 29, 151, 152, 154
Panthera tigris altaica (Siberian tiger), 158
Papouchis, C.M., 178

Paquet, P., 135
Parabuteo unicinctus (Harris hawk), 111
Paramount Farming Company, 128
Pardofelis marmorata, 156
Passerculus sandwichensis beldingi (Belding's savannah sparrow), 127
PATCH model, 251
Peregrine falcon *(Falco peregrinus)*: collision with windows, 110; educational value of, 109; recovery of, 6; urban nesting sites of, 107, 109; urban poisoning of, 110
Phillips, M.K., 178
PHVA (population and habitat viability assessment), 250–251
Pica hudsonia (magpie), 4
Pinus albicaulis (whitebark pine), 161
Poisoning: birds of prey, 109–110, 113; nontarget species, 120, 126–127; San Joaquin kit fox, 120; swift foxes, 120
Polecat *(Mustela putorius)*, 154
Politics: Alaska policy issues, 205–209; ballot initiatives, 203–205; Florida panther conservation and, 187–193; following delisting (overview), 197–198; grizzly bear recovery and, 188; mountain lion conservation and, 223–224, 226–229, 233; overview, 177–178, 266–267; state wildlife commissions, 200–203; wolf management and, 187, 188, 198–199, 205–215, 241–243
Pollution, 5, 160
Population and habitat viability assessment (PHVA), 250–251
Population sinks, 110–111
Porcupine *(Erethizon dorsatum)*, 156
Prairie dog (*Cynomys* spp.), 156, 163
Predator aversion techniques: mountain lions, 233, 234; wolves, 67
Predator control methods: costs of, 127; difficulties with, 68; public opinion of, 67–68, 69, 127; wolves and, 44–45, 53, 57, 58, 59, 60, 61, 63–65, 67–68, 69
Primm, S., 199
Procyon locator. See Raccoon
Pronghorn antelope *(Antilocapra americana)*, 54
Protect Pets and Wildlife (ProPAW), 126–127
Pseudorabies, 83
Puma. *See* Mountain lions
Puma concolor. See Mountain lions
Puma concolor coryi. See Florida panther

Rabies, 83, 124, 157
Raccoon *(Procyon locator)*: cultural inheritance and, 97–98; diet of, 90; highways and, 132; home range size of, 92–93, 95, 97; nuisance issues with, 1, 82, 83; public health issues with, 1, 83–84. *See also* Urban landscapes/skunks, raccoons, coyotes
Raccoon roundworm, 84
Rallus longirostris levipes (California lightfooted clapper rail), 127
Randomization (research), 14
Raven *(Corvus corax)*, 4
Reagan, Ronald, 226
Recovery Plan for Upland Species of the San Joaquin Valley, California, 120–121
Red fox *(Vulpes vulpes)*, 4, 151
Red fox *(Vulpes vulpes necator)*, 121
Red fox *(Vulpes vulpes regalis)*: colonization in California, 121–122; direct competition with kit foxes, 122, 123; disease transmission to kit foxes, 124; displacement of kit foxes, 122–123; endangered prey of, 127; exclusion by coyotes, 122, 126, 127–128; habitat partitioning with kit foxes, 124–125; niche overlap with kit foxes, 123–124; rabies and, 124; use of edge habitats, 122
Red squirrel *(Tamiasciurus spp.)*, 156
Red-tailed hawk *(Buteo jamaicensis)*, 106
Red wolf, 254
Reithrodontomys raviventris (salt marsh harvest mouse), 127
Replication (research): example, 14; with livestock depredation studies, 17–18, 23–24; metareplication, 15
Research: design overview, 14–15. *See also* Livestock depredation research
Riley, S.J., 223
River otter *(Lutra canadensis)*: fur harvesting, 154; highways and, 141
Robel, R.J., 18
Roosevelt, T., 148, 221
Ruediger, B., 79, 142
Rural landscapes/coexistence: overview, 9–11. *See also* Livestock depredation
Rusz, P., 222

Safe harbor program: description, 128; kit fox and, 128, 264
Salt marsh harvest mouse *(Reithrodontomys raviventris)*, 127

Sando, R., 203
Sandstrom, P., 142
San Joaquin kit fox *(Vulpes macrotis mutica)*: conservation/habitat protection, 128; conservation/red fox control, 126–128; conservation summary, 129, 264; current range of, 119–120; description, 119; federal/state listing of, 120; habitat reduction of, 119–120; poisoning of, 120; *Recovery Plan for Upland Species of the San Joaquin Valley, California,* 120–121; as an "umbrella" species, 118–119, 121
San Joaquin kit fox *(Vulpes macrotis mutica)*/coyote: coevolution of, 125–126; coyotes' exclusion of red foxes, 126, 127–128; direct competition with, 125
San Joaquin kit fox *(Vulpes macrotis mutica)*/red fox: direct competition, 122, 123; disease transmission, 124; habitat partitioning/water, 124–125; kit fox displacement, 122–123; niche overlap, 123–124
San Joaquin Valley, California, 118
Sargeant, A.B., 126
Save the Tiger Fund, 266
Schaller, D., 254
Sea otter *(Enhydra lutris)*, xii, 2, 3
Sea otter, southern, 5
Sea urchin *(Strongylocentrotus puratus)*, xii, 3
Servheen, C., 142
Seton, E., 221
Shrader-Frechette, K.S., 191
Siberian tiger *(Panthera tigris altaica)*, 158
Singleton, P.H., 132–133
Sink: population sinks, 110–111; sink effect, 228
Sink packs, 46
Site tenacity, 107
Skunks (Mephitidae family), 151
Skunks, striped *(Mephitis mephitis)*: cultural inheritance and, 98; diet of, 90–91; home range size of, 92, 93–94, 96; nuisance issues with, 82, 83; public health issues with, 83, 84. *See also* Urban landscapes/skunks, raccoons, coyotes
Sloth bear *(Melursus ursinus)*, 151
Snow leopard *(Uncia uncia)*, 154
Sodium cyanide, 127
Source packs, 46
Southern Rockies Wolf Monitoring and Research Team, 249

Southern Rockies Wolf Restoration Project: member groups of, 246; wolf restoration/SRM, 246–248, 256, 257
Southern Rocky Mountains (SRM). *See* Wolf *(Canis lupus)* restoration/Southern Rocky Mountains
Specialists endangerment, 163, 164, 165, 168, 171–172, 173
Species Survival Commission, IUCN, 250
Spectacled bear *(Tremarctos ornatus)*, 151
Squires, 139–140
Stahl, P., 15
Stakeholders: Alaska wolf management, 208–209; Montana wolf management, 213–214; mountain lion conservation, 230–231, 232–233; Wyoming wolf management, 212–213
State wildlife commissions: Alaska wolf management, 205–209, 212; ballot initiatives and, 203; commission membership, 200; creation reasons, 200, 201; criticism of, 201–203; Florida panther conservation and, 189–191; funding and, 201, 214–215, 216; Idaho's Wolf Conservation and Management Plan, 210–211, 214; management paradigm, 201; Montana's Wolf Conservation and Management Plan, 213–214; nongame/environmental interests and, 200, 201–203, 207–208, 209; responsibilities of, 200–201; wolf management, 205–214, 242; Wyoming's Wolf Management Plan, 211–213
Sterna antillarum browni (California least tern), 127
Strongylocentrotus puratus (sea urchin), xii, 3
"Suitability of the Southern Rockies for Wolf Restoration: An Ecological and Social Assessment," 253, 255
Sun bear *(Helarctos malayensis)*, 151
Swainson's hawk *(Buteo swainsoni)*, 108
Sweanor, L.L., 227, 234, 235
Swift fox *(Vulpes velox)*, 120

Tamiasciurus spp. (red squirrel), 156
Taxidea taxus (badger), 5, 144
Teel, T.L., 232
Terborgh, J., 229
Tiger *(Panthera tigris)*, 29, 151, 152, 154
Timm, R.M., 99

Torres, S., 226, 230
Tourism economy: carnivores and, 2, 254–255, 263, 265; wolf restoration in Yellowstone, 2, 254–255, 263
Toxoplasmosis, 83
Trans-Canada Highway, 139, 145
Trapping, 126–127
Tremarctos ornatus (spectacled bear), 151
Trichomonas gallinae, 110
Trichomoniasis, 110–111, 113
"Trophic cascade": description, 3; elk/wolf example of, 3–4
"Truth About Wolves, The"/campaign, 255, 256

Umbrella species: description, 4, 121; Florida panther as, 192; grizzly bear as, 4; San Joaquin kit fox as, 118–119, 121
Uncia uncia (snow leopard), 154
Urban landscapes/birds of prey: aesthetic value of birds, 108; "anthropogenic ecological traps," 110; background, 105; benefits to birds, 107–108; benefits to people, 108–109; birds as nuisance, 111–112; bird selection of site, 106–107; coexistence overview, 114; collision with windows/vehicles, 110, 112–113; diseases and birds, 110–111, 113; diving at people by birds, 111, 113–114; electrocution of birds, 109, 111, 112–113, 263–264; harassment of birds, 111, 113; human education and, 107, 108–109, 113, 114; managing problems for birds, 112–113; managing problems for people, 113–114; poisoning of birds, 109–110, 113; population sinks, 110–111; population sources, 111; problems for birds, 109–111; problems for people, 111–112
Urban landscapes/coexistence: conflicts overview, 77–78; overview, 77–79. *See also* Highways/carnivores
Urban landscapes/skunks, raccoons, coyotes: anthropogenic stimuli and, 98–100; benefits of, 84–85; benefits of mesocarnivores, 84–85; conclusion, 100; conflicts overview, 82–84; conservation and, 84; cultural inheritance, 97–98; diet, 90–92; ecological role of, 84–85; education for public and, 100; management, 98–100; monitoring methods, 87; movements/spatial organization, 92–97;
nuisance issues, 82–83; population density, 87–90; public health issues, 83–84; study area, 85–87; translocation of wildlife, 99; trash reduction/access, 99
Ursus americanus. *See* Black bear
Ursus americanus floridanus (Florida black bear), 155, 187
Ursus arctos. *See* Grizzly bear
Ursus arctos arctos (European brown bear), 163
U.S. Fish and Wildlife Service: best science and, 249; ESA and, 197; Florida panther and, 188–189, 190; investigation of depredation complaints, 55; safe harbor program of, 128; wolf recovery and, 241, 242–243, 247–248, 256

Voight, D.R., 126
Vulpes macrotis mutica. *See* San Joaquin kit fox
Vulpes velox (swift fox), 120
Vulpes vulpes (red fox), 4, 151
Vulpes vulpes necator. *See* Red fox
Vulpes vulpes regalis. *See* Red fox

Walker, R., 142
Waller, J.S., 142
Washington Department of Fish and Wildlife, 224, 232, 233
Weasel (*Mustela* spp.), 144, 151
Weaver, J.C., 135
Wetlands: California loss of, 118; mitigation and, 137
White, P.J., 125
Whitebark pine *(Pinus albicaulis)*, 161
White-tailed deer *(Odocoileus virginianus)*, 54, 85
White-winged dove *(Zenaida asiatica)*, 110
Wilcox, B.A., 137
Wolf *(Canis lupus)*: Alaska wolf management, 205–209; animal rights and, 52, 207; colonization of Montana, 53; costs/benefits of, xii; depredation on pets, 43, 44, 51; dimensions of debate over, 198–199, 206–209; downlisting of, 53; encounters between wolf packs, 43; ESA listing of, 52, 241; gender/conflicts, 31; highways and, 132, 135, 139, 140, 144, 155; persecution/extirpation of, 51–52, 53, 120, 151, 154, 172,

220, 240–241; poisoning of, 120; rural vs. urban people and, 52; status in Canada, 52–53, 68; supporters of, 152; wildlife crossings and, 144

Wolf *(Canis lupus)*/depredations in western North America: animal husbandry practices, 64, 65–66; aversive conditioning, 67; background, 51–54; compensation costs, 58, 60; compensation programs, 53–54, 66; conclusion, 68–70; conflict minimization, 66–68; depredation data, 59, 60, 61, 64; depredation events classification, 56; depredation reporting, 56; discussion, 62–65; fladry use/effectiveness, 69–70; harassment effects, 64; lethal control, 53, 57, 58, 59, 60, 61, 63–65, 67–68, 69; seasonal depredation patterns, 59, 61–62, 65–66; study analysis, 57; study area, 54, 55; study data, 55–56; study methods, 54–57; study results, 57–62; translocation of wolves, 67; USFWS investigation of depredation complaints, 55; U.S. vs. Canada, 56–57; wolf electric shock collars, 67; wolf learning and, 63–64; wolf prey species, 54; wolves expansion/livestock depredation, 57, 58, 59, 60, 62–63, 68–69; wolves killed/livestock depredation, 57, 58, 59, 60, 61; wolves killed/reported, 56–57

Wolf *(Canis lupus)*/depredations in Wisconsin: background, 31–32; compensation payment, 45; conclusion, 46; conflict minimization, 44, 45–46; control operations, 44–45; depredation categories, 31; GIS use, 35; hunting dogs depredation, 31, 35–43, 44, 46; landscape features and, 35, 36, 39–40, 43; livestock depredation, 31, 35–41, 43–44, 45, 46; pack size and, 37–39, 40–41, 43, 44; pack tenure and, 37, 38, 39; predictability and, 44–45, 46; repeated depredation, 36–37, 43; sterilization techniques, 45–46; study analysis, 34–35; study discussion, 41, 43–46; study methods, 32–34; study results, 35–41, 42; summer howl surveys, 32, 33; switch to domestic animals, 31, 36, 43–44; wolf electric shock collars, 45; wolf home range and, 35, 43; wolf pup count and, 37, 38, 44; wolf reactions to dogs, 41, 43, 44

Wolf *(Canis lupus)* management: by Alaska, 205–210, 212; delisting process, 197–198; Idaho's Wolf Conservation and Management Plan, 210–211, 214; Montana's Wolf Conservation and Management Plan, 213–214; questions regarding, 209–210; tourism boycott threats over, 212; wolf debate dimensions, 198–199, 215–216; wolf-related ballot initiatives, 205–206, 207, 208, 210; Wyoming's Wolf Management Plan, 211–213, 215

Wolf *(Canis lupus)* recovery: distinct population segments (DPS) and, 241, 242–243; political actions following, 241–243. *See also* Wolf *(Canis lupus)* management

Wolf *(Canis lupus)* restoration/Southern Rocky Mountains (SRM): background, 240–246; best science and, 249–255; capacity estimates, 252; economics of, 254–255; education/outreach, 248–249, 255–256; education/outreach coalitions, 248–249; habitat suitability study, 250, 251–252; overview, 256–258; population and habitat viability assessment (PHVA), 250–251; public opinion polls, 253–254; research coalitions, 249; restoration promotion coalitions, 246–248; significance of, 244–245; Southern Rockies Wolf Restoration Project, 246–248, 256, 257; SRM description, 243–244; USFWS wolf classification for, 247–248, 256; wolf restoration feasibility report, 252–253

Wolf *(Canis lupus)* restoration/Yellowstone: controversy over, 107; elk/trees and, 2–3; foxes/coyotes and, 4; other animals and, 3–4; population numbers of, 5–6; public comments on, 5; recolonization and, 197; recovery and, 140; reintroduction, 3, 53; tourism economy and, 2, 254–255; *Wolves for Yellowstone?*, 252–253

Wolf Forum for the Southern Rockies, 248–249, 256, 257

Wolf Policy Project, 206

Wolverine *(Gulo gulo)*: highways and, 132, 133, 136, 139, 140, 141; livestock depredation and, 154; population counts of, 5; problem individuals with, 16

World Conservation Union, 168

Wydeven, A., 10

Wyoming Game and Fish Department (WGFD): funding and, 201, 215; mountain

lions and, 224; wolf management by, 211–213, 215
Wyoming's Wolf Management Plan, 211–213

Yaak River Valley, 141
Yellowstone National Park: highways/carnivores and, 138. *See also* Wolf *(Canis lupus)* restoration/Yellowstone

Zenaida asiatica (white-winged dove), 110
Zenaida macroura (mourning dove), 110
Zoonoses, 83

Island Press Board of Directors

Chair
Victor M. Sher, Esq.
Sher & Leff
San Francisco, CA

Vice-Chair
Dane A. Nichols
Washington, DC

Secretary
Carolyn Peachey
Campbell, Peachey & Associates
Washington, DC

Treasurer
Drummond Pike
President
The Tides Foundation
San Francisco, CA

Robert E. Baensch
Director, Center for Publishing
New York University
New York, NY

David C. Cole
President
Aquaterra, Inc.
Washington, VA

Catherine M. Conover
Quercus LLC
Washington, DC

William H. Meadows
President
The Wilderness Society
Washington, DC

Henry Reath
Collectors Reprints
Princeton, NJ

Will Rogers
President
The Trust for Public Land
San Francisco, CA

Charles C. Savitt
President
Island Press
Washington, DC

Susan E. Sechler
Senior Advisor
The German Marshall Fund
Washington, DC

Peter R. Stein
General Partner
LTC Conservation Advisory Services
The Lyme Timber Company
Hanover, NH

Diana Wall, Ph.D.
Director and Professor
Natural Resource Ecology
Laboratory
Colorado State University
Fort Collins, CO

Wren Wirth
Washington, DC